Biogeography

an ecological and evolutionary approach

Biogeography
an ecological and evolutionary approach

by C. BARRY COX PhD, DSc
and PETER D. MOORE PhD
Division of Life Sciences, King's College, London

FIFTH EDITION

OXFORD
BLACKWELL SCIENTIFIC PUBLICATIONS
LONDON EDINBURGH BOSTON
MELBOURNE PARIS BERLIN VIENNA

© 1973, 1976, 1980, 1985, 1993 by
Blackwell Scientific Publications
Editorial Offices:
Osney Mead, Oxford OX2 0EL
25 John Street, London WC1N 2BL
23 Ainslie Place, Edinburgh EH3 6AJ
238 Main Street, Cambridge
  Massachusetts 02142, USA
54 University Street, Carlton,
  Victoria 3053, Australia

Other Editorial Offices:
Librairie Arnette SA
2, rue Casimir-Delavigne
75006 Paris
France

Blackwell Wissenschafts-Verlag
Meinekestrasse 4
D-1000 Berlin 15
Germany

Blackwell MZV
Feldgasse 13
A-1238 Wien
Austria

First published 1973
Second edition 1976
Third edition 1980
Fourth edition 1985
Reprinted 1988, 1989, 1991
Fifth edition 1993

Set by Setrite Typesetters, Hong Kong
Printed and bound in Great Britain
at the University Press, Cambridge

DISTRIBUTORS

Marston Book Services Ltd
PO Box 87
Oxford OX2 0DT
(*Orders*: Tel: 0865 791155
      Fax: 0865 791927
      Telex: 837515)

USA
Blackwell Scientific Publications, Inc.
238 Main Street
Cambridge, MA 02142
(*Orders*: Tel: 800 759-6102
          617 876-7000)

Canada
Oxford University Press
70 Wynford Drive
Don Mills
Ontario M3C 1J9
(*Orders*: Tel: 416 441-2941)

Australia
Blackwell Scientific Publications
Pty Ltd
54 University Street
Carlton, Victoria 3053
(*Orders*: Tel: 03 347-5552)

A catalogue record for this title
is available from the British Library

ISBN 0-632-02967-6

Library of Congress
Cataloging in Publication Data

Cox, C. Barry (Christopher Barry),
  1931— Biogeography: an ecological
  and evolutionary approach/C. Barry
  Cox and Peter D. Moore. — 5th ed.
    p.   cm.
Includes bibliographical references
and index.
ISBN 0-632-02967-6
1. Biogeography. I. Moore,
Peter D. II. Title.
QH84.C65   1993
574.5 — dc20

Contents

Preface to the Fifth Edition

The 20 years that have elapsed since the first edition of this book have seen major changes in biogeography, which are reflected in this new edition. Some of these have involved the appearance of new techniques of analysis, such as the Theory of Island Biogeography, or the varied (and competing) techniques of interpretation of the data of historical biogeography. As well as increasing the range and detail of established types of data, such as range changes during the Ice Ages, some areas of information have increased to a point at which they are now beginning to make a major contribution to our understanding. For example, the use of computer models in palaeoclimatology is now not only allowing us to understand the interactions of known geological or geographical changes in affecting the world's climatic patterns, but also to make predictions as to what changes might have been expected — even if sometimes these predictions do not seem to conform to what appears to have actually happened.

With increasingly detailed information, we are now also beginning to understand the complicated interactions between plate tectonics, sea-levels and ocean currents, climate and evolutionary change, or for example between plate tectonics and the early spread of flowering plants. Increasingly, the prime factors of biogeography can now be seen as, on the one hand plate tectonics, expressing itself as varied changes in the physical world, and on the other hand the response of the biological world, by means of evolutionary change.

But the most fundamental change is probably in our focus of attention. We are now beginning to realize the extent to which the present-day biogeography of almost every part of the world is already permeated and altered by the activities of humankind. In the first edition of this book, a few pages were devoted to biological conservation and to the alteration of habitats by human activity. At that time, there was still little understanding of the depth of the threat that we pose to our planet's biota and climate, and the 'greenhouse effect' was still mainly a concern for horticulturalists rather than a worry for the whole planet. Now, the science of biogeography can be seen as central to our understanding of the interactions between the physical world and the living world, and of our impacts on both of those. If we cannot understand present-day biogeography and its historical roots, we shall have little hope of ensuring that the biogeography of our world in the future is beneficial, rather than disastrous, for all species. That concern with the implications of biogeography for our own activities and future is now central to the approach of this new edition.

Preface to the First Edition

Biogeography lies at the meeting-point of several fields of study. To understand why a particular group of organisms is found in a specific area requires knowledge of the organism's ecological relationships — why it is associated with this soil, this temperature range, or this type of woodland. To understand why it is found in particular areas of a continent may require knowledge of the climatic history, which may have led to the isolation of scattered, relict communities. Finally, to understand why it is found in some continents and not in others involves knowledge of the evolutionary history of the group itself and of the geological histories of the land-masses, as the processes of continental drift transported them across the globe, splitting them asunder or welding them in new patterns.

As the impact of man upon the earth's animals and plants increases dramatically, we have become more aware of the need to control the effects of our social and industrial habits. Biogeographers and conservationists therefore now also need to understand how man's present relationship to his environment has come about, and the ways in which he has altered, often radically, the structure of the ecosystems he inhabits or uses.

Despite biogeography's connections with ecology, geography, geology, evolutionary history and economic anthropology, it is usually studied in conjunction with only one or two of these disciplines. Our intention here has been to provide enough of the basic elements of all these approaches, so that a student who specializes in only one of them may nevertheless be able to understand the way in which all of them interact to produce the apparently bewildering array of patterns of distribution of life.

1 Introduction to biogeography

There is one thing that we all have in common; we all share the same planet. For all of us this is home. For this reason, and also because rising human population and declining resources are placing the earth under greater strain, we are now looking to the scientists who study the earth and its living creatures to advise us how best to manage the planet to ensure its future, and with it our own.

Among the sciences involved in this difficult but vital task is biogeography, the study of living things in a spatial and temporal context. Biogeographers seek to answer such basic questions as why are there so many living things? Why are they distributed in the way that they are? Have they always occupied their current distribution patterns? Is the present activity of human beings affecting these patterns and if so, what are the prospects for the future?

One of the most impressive features of the living world is the sheer diversity of organisms it contains, and one of the main problems facing a biogeographer is how to explain this diversity, and also the reasons for the varying patterns of occurrence of different species over the surface of the planet. Why, for example, is there more than one species of seagull? And why do different species of gull have different patterns of distribution, some being widespread and others very local? Why are there so many different types of grass growing in the same field, all apparently doing precisely the same job? Why are there more species of butterfly in Austria than in Norway? It is the task of the biogeographer not only to answer such specific questions, but also to seek general rules that can account for many such observations, and which will provide a general framework of understanding that can subsequently be used for predictions about the consequences of tampering with the natural world.

One of the world's foremost experts on biological diversity, Edward O. Wilson of Harvard University, has claimed that the diversity of life on earth was greater at the time of the origin of the human species than it had ever been before in the course of earth's history. The arrival and cultural development of our species has evidently had, and continues to have, a profound impact on the world's biogeography, modifying species' ranges and bringing some to extinction. From the point of view of biological diversity (the subject of Chapter 2 of this book), the evolution of humankind was something of a catastrophe, and it would be quite unrealistic to attempt any synthesis of biogeography without taking the human impact into account. Very few species escape the effects of human activity in some aspect of their ecology and distribution. For this reason our species will play an important role in this book, not only in terms of our influence on other species of plant and animal, but also because we too are one species among many, perhaps as many as 30 million other species, and we obey essentially the same rules as the others. The more we can understand about the Hawaiian goose, the oak tree and the dodo, the more we shall appreciate our own position in the order of things.

Because it faces such wide-ranging questions, biogeography must draw upon an extensive range of other disciplines. Explaining biodiversity, for example, involves the understanding of climate patterns over the face of the earth and the way in which the productivity of photosynthetic plants differs with climate and latitude. We must also understand what makes particular habitats desirable to animals and plants; why locations of particular soil chemistry, or moisture levels, or temperature range, or spatial structure, should be especially attractive. Hence, climatology, geology, soil science, physiology, ecology and behavioural sciences must all be invoked to answer such questions.

The time factor is also important. The study of fossils has long made it clear that the diversity of living organisms on earth has not always been the same. New species have arisen and old species have become extinct. Appreciating biodiversity, therefore, involves understanding the mechanisms by which new species arise, and it is an essential part of the biogeographer's work to study the source of novelty, the means by which new species can be generated. We worry, very reasonably, about the current rate of extinction on the planet. How fast can we expect new species to evolve to replace them? How fast can species adapt to cope with the modifications which humankind is making to the climate and the living conditions of the world? This aspect of biogeography will be considered in Chapter 5.

Communities of animals and plants are not like bags of coloured beads, a casual juxtaposition of different types randomly thrown together with no interaction between them. There is a distinct pattern to the way in which individuals react with one another in a population of a single species. Competition with one another may determine which individuals survive and which ones fail, providing fuel for the process of evolution. The species itself usually consists of a series of populations scattered throughout the full range of the species. The range of a species is itself a reflection of how wide is the breadth of tolerance within and between these populations. Populations at or near the edge of a species' distribution may make quite different demands upon the environment from those in the main part of the range. But the distribution limits of some species may be related to quite separate problems, such as the existence of an insuperable barrier, like a mountain range or an ocean. Alternatively, some historical or evolutionary accident may have stranded a species, or even a population of a species in a particular location. Hence, some species' ranges consist of fragmented populations, no longer in contact with one another.

Species also interact with one another, but in a somewhat different way, for they are unlikely to be the same shape or size, or to have exactly the same food requirement or preference for a particular habitat. Such interaction between species can eliminate those that are poorly adapted to cope with the prevailing conditions and select those that can cope not only with the physical environment, but also with one another. This means that assemblages of species can be recognized as communities, showing a degree of cohesion and interdependence, and also sometimes a degree of inertia, or stability, when faced with environmental change. Communities are not fixed entities that behave like whole organisms, but they may change their constitution, losing some species and gaining others as populations respond to changing conditions in their own peculiar ways. Communities themselves change with time.

Communities also interact with the non-living world. Light energy is trapped from the sun and passes from the photosynthetic plants along pathways of consumer organisms who tap energy at different levels along the line, ultimately ending with detritivores and decomposers that take up the residue of unused energy and make their living by degrading the hard-won chemical energy into heat during the process of respiration. The chemicals involved in this energy flow are derived from the rocks and soils, from the rainfall and from the atmosphere, and are used and reused in a cycle of uptake and release as the living components of this organization are spawned, grow, die and decompose. This level of interaction between the living and non-living world is termed the ecosystem, and the concept can be useful in biogeography to help one to understand the way in which the natural world operates and the likely impact of changing any component. Changing energy levels, changing carbon content in the atmosphere, changing quantities of fertilizers in the soil or pollutants in the atmosphere, all have their influence on the function of the entire ecosystem, hence the concept is a useful one for the biogeographer.

These four ways of looking at living things in their environment — individuals, populations, communities and ecosystems — are considered in Chapter 4.

The concept of the ecosystem can be used at almost any scale, from a rotting log right up to the entire earth. Each can be considered as an independent system interacting with other systems and hence with its budget for the energy and materials that it uses. The earth is a somewhat peculiar ecosystem in that it is more isolated from its surroundings than most others, receiving energy from the sun, but being essentially a closed system as far as matter is concerned. One frequently used level of scale at which the ecosystem concept is applied in biogeography is that of the biome, or formation. The division of the earth into certain blocks of vegetation, together with associated animals, is a convenient, if artificial, practice and permits the development of a framework for global mapping of ecosystems and for ease of communication between scientists. Hence, such terms as tropical rain forest, savanna grassland, boreal forest and tundra have become everyday expressions and are widely used and understood.

The classification of the living world using this biogeographical, rather than a taxonomic, approach is particularly useful because it allows the development of general statements concerning the function and the future of these units that would otherwise be difficult. Although the tropical savanna grasslands of Brazil, Kenya, India and Thailand comprise different plant and animal species, they have much in common in terms of their architecture, seasonal growth and productivity, life form of animals and plants, and the ways in which they have been used by human populations. Management practices are similar in different geographic areas, hence much that can be learned from one region can be applied to another. In a book about global biogeography, therefore, it is natural that some space should be devoted to these important units. On the other hand, whole books could easily be (and often have been) devoted to a single biome, so we have to be satisfied with a very brief condensation of the major features of the biomes, emphasizing the problems facing them from human impact in the modern world. Accounts of the major biomes are included in Chapter 4.

A very special case is that of islands. Isolated to varying degrees from the continents, these sites have long been recognized for their importance in the study of biogeography and evolution. The study of the plant and animal life of islands played a very important part in the development of Charles Darwin's ideas concerning natural selection as the driving force of evolution and, more recently, islands have provided a model system from which some valuable concepts have been developed concerning the relationship between immigration and extinction in determining the diversity of life in an area. Such studies can be used in predicting the relationship between species richness and site area, which can be of value in the study and design of nature reserves and protected areas for species conservation. The importance due to islands in biogeography has led us to devote Chapter 6 specifically to this topic.

Not only do the patterns of species differ over the modern globe, but they have also varied in time. The fossil record enables us to look into the past and observe the changing nature of living organisms and their distribution patterns during the passage of time. Our experience of nature represents just one point in a constantly changing mosaic of animals and plants that are responding to an endless course of environmental and climatic change. Biogeography, therefore, can never be a static discipline; it must always be aware of and take account of the modifications that organisms are making to their distribution patterns as conditions favour them or render their life more difficult in a particular region. When conditions change too fast to permit genetic adjustment and evolution, populations, species, and even whole groups of species may pass into extinction. Extinction is not an exception to the norm in evolutionary terms, it is a regular feature of the constantly changing pattern of life.

Understanding biogeography must involve the time perspective, not simply encompassing the decades or centuries by which we measure human history, but covering the millions of years, or even hundreds of millions of years, during which whole groups of plants and animals may rise to prominence and progress to extinction. Within this time scale even the continents have not remained static. Convection forces in the earth's mantle have carried land masses over the surface of the globe, resulting in the fragmentation of supercontinents and occasional continental collisions of great violence, and these movements have had significant repercussions in the distribution patterns of plants and animals. Land mass fragmentation and movement have led to related groups of organisms being widely separated, and collisions have brought unexpected groups close together. So, no account of biogeography can be complete without a consideration of these long-term geological movements. Chapters 7 and 8 provide an overview of these processes and their impact on biogeographical thinking.

Biogeography, then, is concerned with the analysis and explanation of patterns of distribution, and with the understanding of changes in distribution that have taken place in the past and are taking place today. It concerns itself with what units of life, or 'taxa', are found where, and what are the geographical definitions of that 'where' — is it North America, or land between 2000 and 5000 metres altitude, or land between 17°N and 23°S latitude? Once the pattern has been established, two sets of questions arise. There is the internal question: how is the organism adapted to the conditions of life in this area? This in turn generates the opposite, externally directed questions: why does the organism not exist in adjacent areas, and what are the factors (biological or environmental) that prevent it from doing so? These factors can be thought of as forming a barrier that prevents the further spread of the organism.

Sometimes the organism's pattern of distribution is discontinuous or *disjunct*, the species being found in several separate areas. Two different explanations may be offered for this. Firstly, the organism may originally have been present in only one area, and been able to cross the intervening barrier regions to colonize the other areas. Alternatively, the barrier areas may have appeared later, dividing a once-continuous, simple pattern of distribution into separate units. These two explanations are respect-ively termed 'dispersalist' and 'vicariance' (see also p. 237).

In the case of living organisms, all of these questions can be investi-gated in detail, examining all aspects of their ecology and adaptations to life and dispersal. These investigations can extend over several decades and analyse, for example, the effects of a series of dry summers that may cause local ecological changes affecting the distribution of individual species over a few thousand square metres. The wide range of phenomena available for analysis under these circumstances has led to this type of biogeographical investigation being known as 'ecological biogeography'.

Biogeographical changes that have taken place over the last few centuries are similarly well documented, and their causes are well understood. They are also restricted to observable, much-studied phenomena such as climatic change or minor geographical changes in the distribution of land and water. Furthermore, they affected taxa that are comparatively well defined and that can be studied as living organisms. The distributional and climatic data available from cores of lake or deep-sea sediments, or analysis of tree rings, together with radiocarbon dating, have made it possible to extend this ecological biogeography back to several thousands of years before the present.

At a slightly greater remove in time, the biogeographical changes of the Ice Ages can be seen as extrapolations of ecological biogeography. They affected entire biota rather than merely individual species, involved changes in the patterns of distribution and biological zonation over several thousand kilometres, and were caused by regular changes in the earth's orbit or by minor changes in sea levels or currents. Nevertheless, the processes, phenomena and organisms involved are similar in nature to those of today, and the evidence is comparatively abundant.

Understanding even longer-term changes and more ancient patterns of life requires a different approach, which is known as 'historical biogeography'. These changes took place tens or even hundreds of millions of years ago, and involved the splitting, moving or fusion of whole continents, raising new mountain chains or causing the appearance or disappearance of major oceans and seas, with accompanying changes in climate. The biota affected were transported over thousands of kilometres and include groups that are now wholly extinct and are therefore not available for ecological study. The data involved in historical biogeogra-phy are therefore restricted and incomplete, and cannot be approached merely as an extrapolation of the methods used to analyse the phenomena of the more recent past. The different types of data, and the methods that have been proposed for analysing them, form the basis of Chapter 9 of this book.

The last two million years of the earth's history, the subject of

Chapters 10 and 11, have been of particular significance to biogeography for a number of reasons. Within this time period the climate, which has been getting colder for the last 70 million years or so, has become particularly unstable and has entered a series of oscillations in which there is an alternate expansion and contraction of the earth's ice caps. Within these episodes new ice caps formed and shrank over the continental land masses of North America and Eurasia. This glacial/interglacial series of cycles has had a particularly important part to play on the current distribution patterns of plants and animals, not only in the direct effect of changing climates, but also in the alterations it has caused to global sea levels. Rising and falling sea level has periodically resulted in the formation and disruption of land bridges linking islands to land masses and even continents to one another, as in the case of eastern Asia and Alaska. Modern distribution patterns of organisms often provide examples of stranded populations and strangely vacant islands. Since many of the species that now exist have been present for much of the last million years or so, it has proved easier to reconstruct the detailed ecological changes of this ultimate stage in the earth's geological history than has proved possible for earlier stages.

A further reason why this recent period has a particular interest for the biogeographer is the fact that our own species has evolved and come to prominence within the last two million years. The emergence of *Homo sapiens* from the position of a social ape of the savannas to its present position of global dominance has an important message for modern humans. The picture that develops from a biogeographical approach to human development is one of clever manipulation of the environment, deflecting energy from other parts of the natural food webs into the support of one species. This is not unknown in the animal kingdom — some ants herd greenflies and grow fungi in gardens, but in the case of our own species the development of agriculture and later industrial processes has created a set of conditions in the modern world that we find it hard to conceive. It is hard to grasp that one species has achieved such an impact that the massive geochemical and climatic processes of the planet are now affected. In 1959, for example, the following statement could be found in the journal *New Scientist*, 'Nature's carbon cycles are so vast that there seem few grounds for believing man will upset the balance'. Thirty years on, such complacency seems incredible.

The study of biogeography has particular relevance to some of these problems because an understanding of the processes and the factors that have influenced the successes and the failures of other species may well assist in the amelioration of our present condition. Such knowledge may help to underline the necessity for conserving those resources that yet remain to us, so that we may use them wisely. We have much to learn from the Hawaiian goose — a bird that approached the very brink of extinction, yet recovered, and also from the oak trees — a group of species that remains resilient after a long and successful evolutionary history and we must not forget the dodo.

2 Biodiversity

The most obvious questions are often those that are the most difficult to answer. Perhaps the most obvious question to pose to a biologist is 'How many different types of living thing are there?' But the answer to this question is still almost a matter of guesswork. One thing is sure: biologists have as yet succeeded in naming and describing only a very small proportion of the total species present. Edward O. Wilson [1] considers that about 1·4 million species of plant, animal and microbe species are currently known, but other estimates claim 1·7 million. The confusion may seem surprising to non-biologists, but the wealth of information and species makes it difficult to be sure that all of the species described are valid and are not duplicates. It is quite possible that some species have been given separate names as a result of different scientists describing them in different parts of their geographical range. Some 'species' may actually consist of several closely related and similar true species, so this can lead to further confusion. All biologists are agreed, however, that the true number of species still living on the earth far exceeds the number currently described.

A minimal value for the true species number would be about four million, but the tropical ecologist Terry L. Erwin [2] has proposed that the total is far greater than this, perhaps as high as 30 million for tropical insects alone. He came to this conclusion as a result of the study of beetles on a single tree species, *Luehea seemannii*, in Panama, which he sampled by 'fogging'. This is an efficient modern technique for stunning the insects in a canopy by smoking them with an insecticide. The dazed insects fall from the tree and are collected in trays placed beneath the canopy. Erwin examined just 19 individual trees of *L. seemannii* tree species in the Panamanian forests and managed to obtain 1200 species of beetles alone from this analysis. This is not entirely surprising, since beetles are extraordinarily successful insects and may comprise as much as 25 per cent of the total number of species of living organisms. But this study does illustrate the remarkably rich condition of the tropical forest.

From these data Erwin made a number of assumptions about the proportions of different organisms in the forest, and he came to the conclusion that, if this number of beetles were representative, then one might predict a total of 30 million species of insect on earth. The uncertainty of many of the assumptions he makes, however, should make us very cautious in accepting this figure uncritically. Other entomologists, such as Nigel Stork and Kevin Gaston from London's Natural

History Museum [3], have checked Erwin's estimates using data from studies in the tropical forests of Borneo. Stork has generated estimates ranging from 10 million to 80 million for the arthropods (a group of invertebrate animals including the insects). Another independent estimate [4] supports the lower end of this scale, placing tropical arthropods at six to nine million. But the general figures are beginning to find some measure of agreement. No-one has yet been brave enough to attempt an estimate of the diversity of microbes, the bacteria, fungi, viruses, etc. Of the bacteria, for example, only about 3000 species have so far been described, so much remains to be done in this area. Among the fungi, 69 000 have been described, but David Hawksworth of the Royal Botanic Gardens, England, believes that the true total could be around 1·6 million species [5].

A further complication in determining the true number of species present on earth is the rate at which that number is being eroded, mainly as a result of human activity. Wilson has calculated that the loss of species from the tropical forest area alone could be as high as 6000 species per year. This amounts to 17 species each day, and the tropical forests cover only 6 per cent of the land surface area of the earth, so the rate of extinction will be much higher if we consider other vegetation types.

The known species of living organisms are best classified into five major kingdoms [6], of which two, the prokaryota and akaryotic protists, are unicellular. Currently these two groups account for only 5 per cent of known organisms, which is undoubtedly a gross underestimate of their true proportion. The fungi and the plants constitute a further two kingdoms and represent about 22 per cent of known species. This leaves the animal kingdom, which currently accounts for over 70 per cent of known organisms. The imbalance in the proportions of microbes and animals undoubtedly reflects the way in which the attention of researchers has so far been concentrated on the animals rather than any real preponderance of the animal kingdom.

Gradients of diversity

When we look at the way in which species are distributed over the land surface of the planet we find that they are far from even. The tropics contain many more species than an equivalent area of the higher latitudes. This seems to be true for many different groups of animals and plants, as can be seen from Fig. 2.1, which illustrates the number of breeding birds found in various Central and North American countries and states. The tropical country Panama, only 500 miles (800 km) north of the Equator and a close neighbour of Costa Rica, has 667 species, three

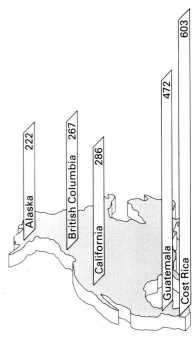

Fig. 2.1 Numbers of breeding bird species in different parts of Central and North America.

times the number of breeding birds found in Alaska, despite the much greater area of Alaska.

A similar pattern is seen in the number of mammal species at different latitudes in North America (Fig. 2.2). If we consider only forest areas from southern Alaska (65°N) in the north, through Michigan (42°N) into the tropical forest of Panama (9°N), these have 15, 35 and 70 mammalian species respectively. Breaking the mammals down into their component groups by diet (see Fig. 2.3) and by taxonomy [7], we find that the bats account for a large part of the difference between the three locations. Moving from the north, one-third of the increase in species from Alaska to Michigan is due to the larger number of bats, and so also is two-thirds of the increase between Michigan and Panama. Yet more information becomes available if we consider diet. Much of the

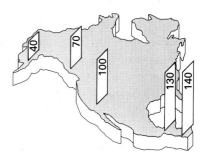

Fig. 2.2 Numbers of mammal species in different parts of Central and North America.

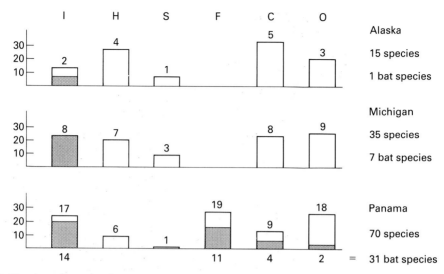

Fig. 2.3 Numbers of mammals in forest sites at three different latitudes in Central and North America. The heights of the columns indicate the percentage of the total mammal species that are insectivorous (I), herbivorous (H), seed-eating (S), fruit-eating (F), carnivorous (C), or omnivorous (O). The number above each column shows the total number of species with that diet. The shaded parts indicate the proportion of bat species, and the numbers below the Panama columns show the number of bat species with that particular diet. From the data of Fleming [7].

tropical diversity is due to the greater predominance of a fruit-eating way of life and to the greater number of insectivores, many of which are eating insects that in turn feed on the fruits of the forest. So diet is evidently an important aspect of the diversity gradient found among animals.

Diet may be important for animals, but what of plants? They also show a general trend towards increasing diversity in the tropics, but they do not vary in their diet since they all need sunlight energy for their photosynthesis. The relationship between plant species richness and latitude, however, is not at all a simple one. David Currie and Viviane Paquin of the University of Ottawa have constructed a map of the richness of tree species across North America [8], and this is shown in Fig. 2.4. From this it can be seen that the contours of richness do not simply follow the lines of latitude, especially in the areas south of Canada. Patches of low diversity occur in the mid-west and exceptionally high diversity of trees in the south-east. When they examined the possible environmental factors that correlated with this pattern, the one which correlated most closely was the sum of evaporation (directly from the ground) and transpiration (from the surface of vegetation) combined to give a value for the loss of water from the land surface

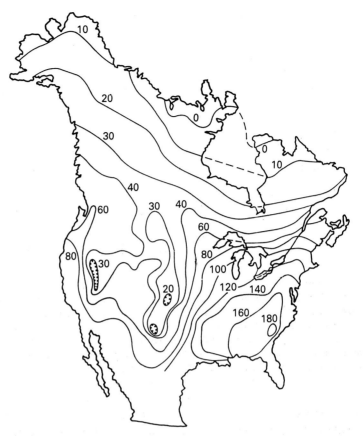

Fig. 2.4 Number of tree species (i.e. any woody plant over 3 m in height) found in different parts of North America. The contours indicate areas where particular numbers of tree species were recorded within large-scale quadrats (mean area of 70 000 square kilometers). Data from Currie and Paquin [8].

(evapotranspiration). Evidently, those regions with the highest evapotranspiration were able to support the highest diversity of tree species. But evapotranspiration also correlates closely with the potential productivity of a region (the amount of plant material that accumulates by photosynthesis in a given area in a given time), so perhaps plant diversity is essentially determined by how much photosynthesis can be carried out in a given site. Figure 2.5 shows this relationship between primary production and tree species richness, and it can be seen that there is indeed a good correlation between the two.

In general, the equatorial regions are the areas in which highest productivity is possible because of the prevailing climate, which is hot, wet and relatively free from seasonal variation. Figure 2.6 illustrates this by displaying the world distribution of primary productivity. From

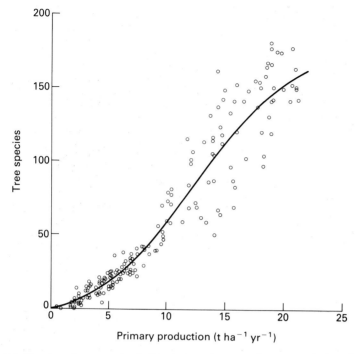

Fig. 2.5 Number of tree species in North American sites (see Fig. 2.4) plotted against the primary productivity of those sites. A distinct positive relationship can be observed. Data from Currie and Paquin [8].

this map it can be seen that very high productivity is concentrated in the equatorial belt and that this drops off as one moves towards higher latitudes. The picture is complicated by the arid belt in northern Africa and central Asia, of which more will be said in due course, but the general trend is decreasing productivity at higher latitudes. An examination of the North American part of this map shows a good correlation with the tree richness map (Fig. 2.4), especially with regard to the high diversity of tree species and high productivity in the south-eastern United States. One theory suggests, therefore, that the richness of plant and animal species in an area is dependent on how much energy is captured by the vegetation.

But this may not be the whole story. Where conditions are most suitable for plant growth, that is where temperatures are relatively high and uniform and where there is an ample supply of water, one usually finds large masses of vegetation. This leads to a complex structure in the layers of plant material. In a tropical rain forest, for example, a very large quantity of plant material builds up above the surface of the ground. There is also a large mass of material developed below ground as root tissues, but this is less apparent. Careful analysis of the above-

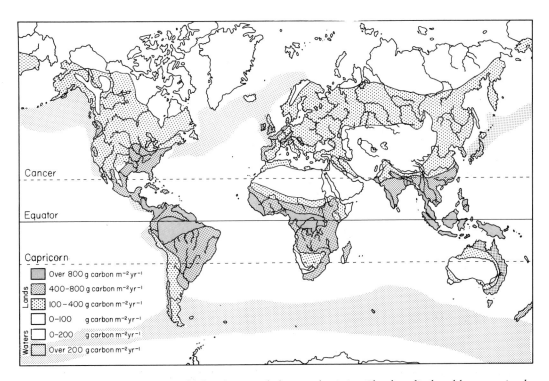

Fig. 2.6 World distribution of plant productivity. The data displayed here are simply estimates of the amount of organic dry matter that accumulates during a single growing season. Full adjustments for the losses due to animal consumption and the gains due to root production have not been made. Map compiled by H. Leith.

ground material reveals that it is arranged in a series of layers, the precise number of layers varying with the age and the nature of the forest. The arrangement of the 'biomass' of the vegetation into layered forms is termed its 'structure' (as opposed to its 'composition', which refers to the species of organisms forming the community). Structure is essentially the architecture of vegetation and, as in the case of some tropical forests, can be extremely complex. Figure 2.7 shows a profile of a mature flood plain tropical forest in Amazonia [9] expressed in terms of the percentage cover of leaves at different heights above the ground. There are three clear peaks in leaf cover at heights of approximately 3, 6 and 30 metres above the ground, and the very highest layer, at 50 metres, corresponds to the very tall emergent trees that stand clear of the main canopy and form an open layer of their own. So this site contains essentially four layers of canopy.

Forests in temperate lands are simpler, often with just two canopy layers, so they have much less complex architecture. Structure, however, has a strong influence on the animal life inhabiting a site. It forms the spatial environment within which an animal feeds, moves around, shel-

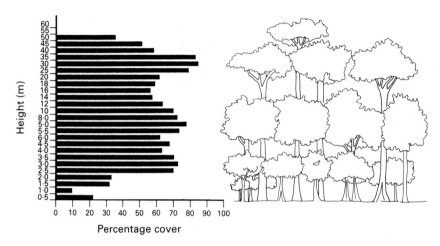

Height (m)

Percentage cover

Fig. 2.7 Profile of a tropical rain forest with the percentage leaf canopy cover recorded at different heights above the ground. Note the stratification of the leaf cover into distinct layers. See Fig. 2.14 for further details of canopy structure development in rain forest. From Terborgh and Petren [9].

ters, lives and breeds. It even affects the climate on a very local level by influencing light intensity, humidity and both the range and extremes of temperature. Figure 2.8 shows a profile through an area of grassland vegetation that has a very simple structure, and it can be seen that the ground level has a very different microclimate from that experienced in the upper canopy. Wind speeds are lower, temperatures are lower during the day (but warmer at night) and the relative humidity is much greater. The complexity of microclimate is closely related to the complexity of structure in vegetation and, generally speaking, the more complex the structure of vegetation the more species of animal are able to make a living there. This is illustrated in Fig. 2.9, which relates the number of bird species found in woodland habitats to the number of leaf canopy layers that can be detected. The high plant biomass of the tropics leads to a greater spatial complexity in the environment, and this will lead to a higher potential for diversity in the living things that can occupy the region.

There is one extension to this line of argument that is worth pursuing. Complexity, or the conception of complexity, depends upon the size of the observer. It was stated above that grassland has a relatively simple structure, but this is only the case if one views it from a human perspective. From an ant's point of view, on the other hand, a grassland environment may be highly complex. For this reason an area of grassland offers a home to far more ants than it does to humans, cows or bison. The earth as a whole can support far more small creatures than it can large ones. This is illustrated by Fig. 2.10, assembled by Robert M. May of Oxford University [10], showing the relationship between the number

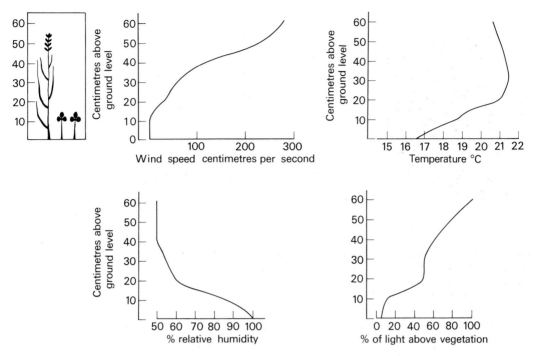

Fig. 2.8 Diagram showing the structure of grassland vegetation and the effect this has upon the microclimate of the habitat.

of species of terrestrial (land) animals and their respective body sizes. The general picture is one of a greater abundance of small species than large ones. The fact that there appears to be a drop in species richness when dealing with extremely small organisms probably reflects our ignorance of these creatures rather than any real decrease in numbers.

The conclusion is, therefore, that the richness of the tropics, especially as far as animal life is concerned, may be a consequence not simply of the high productivity of these latitudes, but also of their great structural complexity resulting from their high biomass, which can support many species of small animal.

If we examine the distribution patterns of the various families of flowering plants, we find that they also have a tendency to centre on the tropics. Very roughly, about 30 per cent of flowering plant families are widespread in distribution, about 20 per cent mainly temperate and about 50 per cent mainly tropical. This has suggested to some research workers that the tropics have been a centre for the evolution of many of the flowering plant (angiosperm) groups. This proposal can be examined by looking at the fossil record to see whether the tropics have always been richer than the temperate latitudes. This approach has been attempted by Peter Crane and Scott Lidgard of the Field Museum of

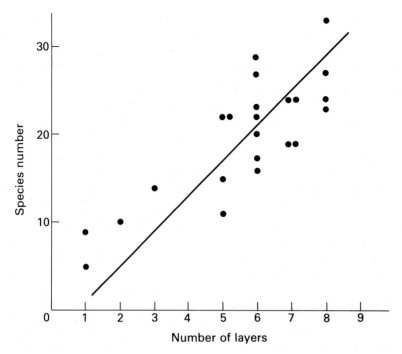

Fig. 2.9 Graph showing the relationship between the number of bird species and the number of layers in the vegetation stratification. Data from Blondell [17].

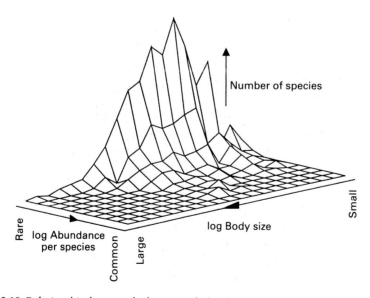

Fig. 2.10 Relationship between body size and abundance among beetle species. The vertical axis represents the number of species of beetle having the body size and abundance indicated on the horizontal axes. 'Small' for a beetle is 0.5 mm and 'large' is 30 mm. From May [10]. Copyright 1988 by the AAAS.

Natural History in Chicago [11], and some of their results are represented in Fig. 2.11. This shows an analysis of fossil plant material covering the period 145 million years ago to 65 million years ago, and depicts the importance of the angiosperms in different latitudes during this period of time. From this it is clear that the flowering plants first rose to some prominence in the tropics, and that their predominance in the tropics was maintained throughout this period of time as they gained importance in the plant kingdom. The latitudinal gradient in diversity, as far as the flowering plants are concerned, goes back a very long way; right back, in fact, to the evolutionary origins of the group.

Arguing from past history to account for the diversity of the tropics with respect to higher latitudes, raises the question of changing global climate with time. It is generally accepted that the earth's climate has been constantly changing, and that over the past two million years it has been considerably colder than has been the case for the previous 300 million years. As a consequence, the high latitudes have been disrupted by the development of glaciers. The effects of these changes on the biogeographical patterns of plants and animals will be considered later, but it is evident that the most severe disruption, in the form of ice masses that have spread and destroyed all vegetation over major areas, has occurred largely in the high latitudes, and the tropics have conse-

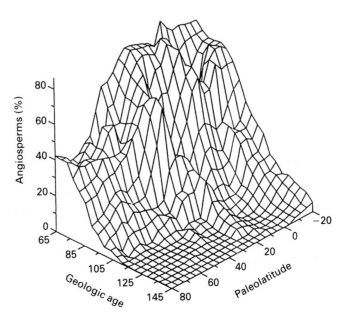

Fig. 2.11 Estimated percentage representation of flowering plants (angiosperms) at different times in geological history and at different latitudes. Angiosperms have always been most abundant in the low-latitude (tropical) regions, suggesting that they originated in this part of the earth. From Crane and Lidgard [11].

quently been subjected to less obvious climatic stress. This idea of a climatically stable tropical belt, if it is indeed true, could account for some of the diversity still found in the tropics — the plants and animals could be a relic feature from a former age. But has the tropical region been climatically more stable than the temperate region? This subject will be examined in greater detail later in the book, but the general conclusion that has emerged, particularly from the studies of Paul Colinvaux of Duke University, North Carolina [12], is that the tropical lowlands of Amazonia have also been considerably colder during recent times (the last million years or so) and that the tropical forests could not have remained intact over the whole of their current range. The rain forest has undoubtedly been restricted in the altitudinal range it was able to cover, and has also been at least partially fragmented as a result of cold and drought during 'glacial' periods.

The equatorial forests, however, have still had to endure less disturbance than their temperate counterparts, and some areas have probably maintained themselves in a forested form throughout the period of stress, even though their composition and structure may well have changed. If they were fragmented, this could have actually assisted in the progress of evolution and diversification, for isolated populations, as we shall see in a subsequent chapter, may diverge in their evolution and form separate species that fail to interbreed when brought into contact once more. The impact of climatic change in fragmenting the tropical forests may have added to, rather than subtracted from, their diversity.

Diversity in time

Some aspects of the spatial gradient in diversity that we find over the land areas of the earth may be explicable in terms of their persistence, even if only as fragments. The study of changing species diversity in time is clearly an important aspect of understanding why there are so many different types of plant and animal, and why they may be concentrated in certain parts of the world. Studies over long periods of time, tens or hundreds of millions of years, can be very informative, but such studies suffer from the disadvantage that it is often difficult to determine how many species there were within a fossil group. One can, however, examine changes in the species composition of a habitat over a short period, decades or centuries. Changes over periods of time of this order are termed successions, especially if they follow a predictable and directional course of development.

A simple example is the invasion of vegetation following the retreat of glaciers in Alaska [13]. Warmer conditions cause the melting of ice, and the ice front gradually recedes, leaving bare rock surfaces and crushed rock fragments in sheltered pockets and crevices. Such primitive soils may be rich in some of the elements needed for plant growth such as potassium and calcium, but are poor in organic matter, so they have a very limited capacity for water retention, poor microbial populations, little structure and low levels of nitrogen. A plant that can grow even under these stressed conditions is the Sitka alder (*Alnus sinuata*). This is a low-growing bushy tree that owes its success in part to its association with a bacterium that grows in association with its roots. This microbe forms colonies in swollen nodules on the alder's roots and is able to take nitrogen from the atmosphere and convert it to ammonium compounds that can subsequently be used (together with materials derived from alder photosynthesis) to build up proteins. So the alder manages perfectly well despite the low levels of nitrate in the soil. In fact, because of the gradual death of roots and the return of litter to the soil from the alder, the growth of the tree increases the amount of nitrate in the soil and thus fertilizes it. But this very process of modifying the soil environment eventually proves the downfall of the alder because it permits the invasion of other, less highly adapted plants, among them the Sitka spruce (*Picea sitchensis*). After about 80 years the Sitka spruce trees, which are more robust and faster-growing than the alder shrubs, assume dominance in the vegetation and begin to shade out the pioneer alders. Thus, by their very existence at the site the alders have effectively sealed their own fate and made the next step in the succession inevitable. This driving mechanism that underlies the successional process is termed facilitation, and it ensures a progressive development within the vegetation.

The course of succession also leads to an accumulation of biomass during the course of time. This is shown in Fig. 2.12, in which the biomass of the major tree species can be seen increasing over the course of 200 years of a succession in Glacier Bay, Alaska. Alder, poplar (*Populus*) and willow (*Salix*) are replaced by Sitka spruce and hemlock (*Tsuga*), and while alder only achieved a maximum biomass of about 50 tonnes per hectare, the spruce/hemlock forest grows to a biomass of over 300 t/ha. Such an increase in biomass naturally involves the development of a more complex canopy structure and, as we have seen, the diversity of animal species closely follows the increase in structural complexity in vegetation. A gap in a tropical rain forest created by the death of an old tree or by wind damage, for example, gradually becomes filled by the invasion and growth of new vegetation [9]. This operates like a succession in miniature, as is shown in Fig. 2.13. The development in structural complexity can be seen in Fig. 2.14, in which plants gradually replace one another and produce successively more complex patterns of canopy cover. This in turn leads to an increasing diversity in the course of succession.

The word 'diversity' should be used with caution, for it can represent

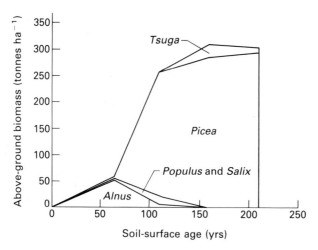

Fig. 2.12 Change in major species composition (expressed in terms of biomass, i.e. above-ground dry weight) during the development of forest following ice retreat in Alaska. The dominance of one species, spruce (*Picea*), is established as the biomass increases with successional development of the plant community. From Bormann and Sidle [13].

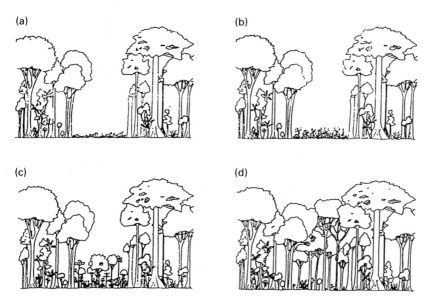

Fig. 2.13 Gaps in forests, created by the death of old trees, or minor catastrophes such as wind-blow or fire, are healed by the regrowth of young trees, often passing through a succession of different species. Habitat heterogeneity and canopy complexity is an outcome of this process.

different things in different people's minds. It conveys the idea of the number of species per unit area of ground, but this is not an adequate definition of diversity. One should really call this 'species richness'.

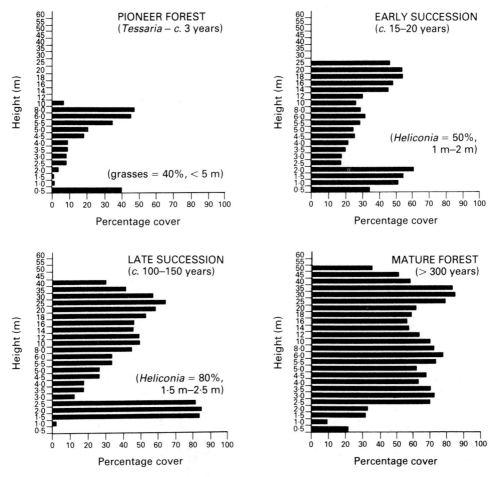

Fig. 2.14 Canopy profiles of a rain forest during the course of successional development over the course of about 300 years. The canopy structure is expressed in terms of leaf percentage cover at different heights above the ground. From Terborgh and Petren [9].

Diversity also involves the way in which the number of animals (or their biomass) is allocated among the species present. A community, for example, that contains 100 individuals belonging to 10 species could have 91 individuals belonging to one species and only one each for all the others. This is a less diverse community than one in which there are 10 individuals belonging to each species. Yet both communities have the same species richness.

In the case of successions, both richness and diversity tend to increase with age, especially for the animal component, but this may not be the case with plants. Later stages in succession, as in the case of the alder/spruce sequence, may become dominated by a few large-bodied species, which effectively reduces plant diversity. In the early stages of succession, however, increasing diversity seems to hold generally true for both plants and animals. This can be illustrated by reference to Fig. 2.15, which shows the number of plant species and their area of cover during the course of an 'old field' succession. This is the development of vegetation following the abandonment of agricultural land and its reversion to woodland. The data shown in the diagram are derived from the work of F.A. Bazzaz [15], who studied an old field succession in Illinois. It covers a period of 40 years and illustrates the increasing number of plant species present (richness) and also a general flattening of the bars, showing that fewer species are dominating the community and more species are occupying a fairer share of the available space. This is a graphic way of expressing the concept of diversity, which can clearly be seen to be increasing through successional time in this instance. There is no strong indication of dominance here, although in later stages one species is beginning to account for a large proportion of the total vegetation cover. (Note that the cover values are expressed as logarithms, which tends to make such dominance less evident.)

The final, mature stage in a succession is termed the 'climax', and represents an equilibrium state. But the achievement of overall equilibrium does not mean that the community is static. Individual trees will become senile and die, minor catastrophes such as wind blow and fire may cause openings to develop within the canopy, and these gaps, as we have seen, become occupied by small-scale successional developments. Some members of pioneer species groups will find new opportunities to survive for a short while and re-establish themselves within the gap, eventually to be replaced by more persistent species.

In the north-eastern United States a typical sequence in the hardwood forest, following the fall of a mature beech tree, is yellow birch invasion, followed by sugar maple and eventually the regrowth of beech [16]. But since this cyclic process of forest healing is taking place wherever a gap has resulted, the 'climax' forest actually consists of a mosaic of patches all in different stages of recovery, together with some patches of mature beech. So the climax vegetation is actually a collection of different-aged patches. This, in fact, adds to the diversity of the whole system, for the vegetation is not uniform but extremely heterogeneous and many of the species that would be lost from an area, if successional development had effectively ceased, are still present in some of the forest openings. Thus the complexity of the time element in vegetation development allows even more species to be packed into a given area.

If we return to the question of why the tropics, and tropical forests in particular, are so rich in species, then we have now established a further means by which high diversity can be maintained. The forest is constantly undergoing disturbance from storms, local fire and the meandering and flooding of rivers. All of these leave the forest in a state

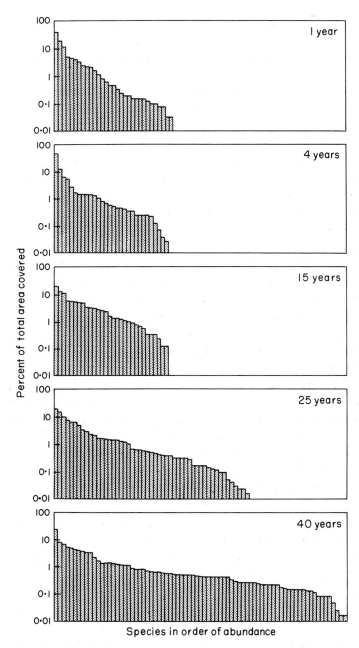

Fig. 2.15 Increasing species diversity in an old field succession in Illinois. Species are ranked in order of abundance, the latter being expressed as percentage area covered on a log scale. Data from Bazzaz [15] (after May [14]).

of turbulence and active regeneration that contributes to the diversity of the whole.

Explaining the global patterns of biodiversity is a process that requires

a consideration of many factors. Some of these
here, but further understanding of the subject den
many other aspects of biogeography. How do ma
occupy the same habitat? What factors limit the
individual species? How have such ranges changed
the earth's history? How do new species evolve an
in particular ways? These are some of the question
if the complex issue of biodiversity is to be furth

References

1 Wilson, E.O. (1988) *Biodiversity*. National Academy Pro
2 Erwin, T.L. (1983) Beetles and other insects of tropical fc
  Brazil, sampled by insecticidal fogging. In *Tropical R
  Management (eds S.L. Sutton, T.C. Whitmore & A.(
  Scientific Publications, Oxford, pp. 59–75.
3 Stork, N. & Gaston, K. (1990) Counting species one by one. *New Scientist*
  127(1729),43–47.
4 Thomas, C.D. (1990) Fewer species. *Nature* **347**,237.
5 May, R.M. (1991) A fondness for fungi. *Nature* **352**,475–476.
6 Margulis, L. & Schwartz, K.V. (1987) *Five Kingdoms*, 2nd edn. W.H. Freeman,
  Oxford.
7 Fleming, T.H. (1973) Numbers of mammal species in North and Central American
  forest communities. *Ecology* **54**,555–563.
8 Currie, D.J. & Paquin, V. (1987) Large-scale biogeographical patterns of species
  richness of trees. *Nature* **329**,326–327.
9 Terborgh, J. & Petren, K. (1991) Development of habitat structure through succession in an Amazonian floodplain forest. In *Habitat Structure: The Physical
  Arrangement of Objects in Space* (eds S.S. Bell, E.D. McCoy & H.R. Mushinsky).
  Chapman & Hall, London, pp. 28–46.
10 May, R.M. (1988) How many species are there on earth? *Science* **241**,1441–1449.
11 Crane, P.R. & Lidgard, S. (1989) Angiosperm diversification and paleolatitudinal
  gradients in Cretaceous floristic diversity. *Science* **246**,675–678.
12 Colinvaux, P.A. (1989) The past and future Amazon. *Scientific American*
  260(5),68–74.
13 Bormann, B.T. & Sidle, R.C. (1990) Changes in productivity and distribution of
  nutrients in a chronosequence at Glacier Bay National Park, Alaska. *Journal of
  Ecology* **78**,561–578.
14 May, R.M. (1978) The evolution of ecological systems. *Scientific American*
  238(3),118–133.
15 Bazzaz, F.A. (1975) Plant species diversity in old field successional ecosystems in
  southern Illinois. *Ecology* **56**,485–488.
16 Forcier, L.K. (1975) Reproductive strategies in the co-occurrence of climax tree
  species. *Science* **189**,808–810.
17 Blondell, J. (1979) *Biogeographie et Ecologie*. Masson, Paris.

3 Patterns of distribution

The basic units with which biologists have to operate are individual organisms, whether animals, plants or microbes. In the majority of cases these individuals can be sorted into like groups which we call species. Species are thus reasonably distinguishable groups of organisms within which interbreeding can occur and to which it is normally confined (see p. 14). Having said this, however, one must bear in mind that there may be considerable variation within any given species, both in visual features, such as form, size and colour, and in physiological and biochemical features that may affect the preferred environment, food or climatic tolerance [1]. Such variations may become so distinctive that they justify the erection of subspecies or races as convenient systems for the classification of individuals.

Species are classified into higher units, such as genera and families, again according to the degree of similarity encountered, but such divisions are not always clear and they may not always reflect adequately the degrees of difference or similarity that actually occur between organisms. For example, some biologists consider that the genetic constitution of the chimpanzee is sufficiently similar to that of our human species to warrant our both being placed in the same genus. Such controversy serves to illustrate that the classification of living organisms is by no means cut and dried, and is essentially a convenient system that taxonomists have adopted and are constantly modifying as new information becomes available.

When we examine the geographical distributions of species of organisms we find that there are effectively no two species that are identical in their ranges. Some correspond fairly closely, but others differ totally. When we use terms like 'distribution' and 'range', we must also be careful about the spatial scale we are considering. Two species may be widespread within a given geographical area, such as the British Isles, and yet occupy different types of habitats (such as woodland or grassland). Even within a habitat, species may occupy different microhabitats. In a New Zealand forest, for example, one may find both the brown kiwi (*Apteryx australis*) and the fantail (*Rhipidura fuliginosa*), a kind of flycatcher. But they occupy different microhabitats, for the former is confined to the forest floor whereas the latter nests in canopy branches. So scale is an important consideration when studying distribution patterns [2].

Limits of distribution

Whether the areas of a species' distribution are considered on a geographical, habitat, or microhabitat scale, they are surrounded by areas where the species cannot maintain a population because different physical conditions or lack of food resources will not permit survival. These areas can be viewed as *barriers* that must be crossed by the species if it is to disperse to other favourable, but as yet uncolonized, places — much as the European settlers had to cross ocean barriers to colonize North America or Australia. Any climatic or topographic factor, or combination of factors, may provide a barrier to the distribution of an organism. For example, the problems of locomotion or of obtaining oxygen and food are quite different in water and air. As a result, organisms that are adapted for life on land are unable to cross oceans: their eventual death will be due, in varying proportions, to drowning, to starvation, to exhaustion and to lack of fresh water to drink. Similarly, land is a barrier to organisms that are adapted to life in sea or fresh water, because they require supplies of oxygen dissolved in water rather than as an atmospheric gas, and because they desiccate rapidly in air. Mountain ranges, too, form effective barriers to dispersal because they present extremes of cold too great for many organisms. The amount of rainfall, the rate of evaporation of water from the soil surface, and light intensity are all critical factors limiting the distribution of most green rooted plants. But in all these cases, and in most others, the ultimate barriers are not the hostile factors of the environment but the species' own physiology, which has become adapted to a limited range of environmental conditions. In its distribution a species is therefore the prisoner of its own evolutionary history.

Take the palms, for example. Figure 3.1 shows the global distribution of this plant family, and it can be seen that members are found in all areas of the tropics and in many subtropical regions too. When one comes to the temperate areas, however, such as Europe, there are very few species of palm that can be regarded as native. Indeed, there are only two truly native palms in Europe. The one called *Chamaerops humilis*, is a very small species which grows in sandy soils in southern Spain and Portugal, eastwards to Malta (Fig. 3.2). So a family which is extremely successful and widespread in the tropics has failed to achieve similar success in the temperate regions. The real problem with the palms is the way they grow: they have only a single growing point at the apex of their upright stems, and if this is damaged by frost then the whole stem perishes. This weakness has even limited the use of palms as domesticated, crop plants, for species such as the date palm cannot be grown in areas with frequent frosts. So even in the deserts of northern Iran, the date palm (*Phoenix dactylifera*) is a rare sight because of the

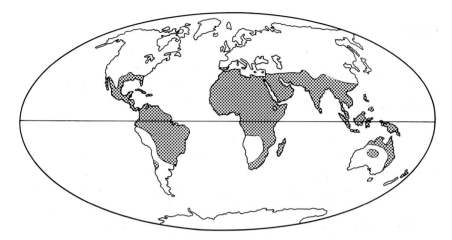

Fig. 3.1 World distribution map of the palm family (Palmae), a pantropical family of plants.

Fig. 3.2 The dwarf palm *Chamaerops humilis,* one of the two native palms found in Europe.

intense cold in these high-altitude drylands during winter (Fig. 3.3). Perhaps the most successful palms are the *Serenoa* species, which reach 30°N in the United States, and *Trachycarpus*, which attains an altitude of 2400 m in the Himalayas. But the family as a whole is limited geographically by its frost sensitivity [3].

Some plants and animals are confined in their distribution to the areas in which they evolved; these are said to be *endemic* to that region. Their confinement may be due to physical barriers to dispersal, as in

Fig. 3.3 The date palm, *Phoenix dactylifera*, at its most northerly site in the Great Kavir Desert of Iran.

the case of many island faunas and floras, or to the fact that they have only recently evolved and have not yet had time to spread from their centres of origin.

At the habitat level the microhabitats of organisms are surrounded by areas of small-scale variation of physical conditions, or *microclimates* — similar to, but on a much smaller scale than, geographical variations in climate — and of food distribution. These form barriers restricting species to their microhabitats. The insects that live in rotting logs, for instance, are adapted by their evolution to a microhabitat with a high water content and relatively constant temperatures. The logs provide the soft woody materials and micro-organisms the insects need for food, and also give good protection from predators. Around the logs are areas with fewer, or none, of these desirable qualities and, for many of the animals, attempts to leave their microhabitat would result in death by desiccation, starvation, or predation.

Overcoming the barriers

Nevertheless, a few inhabitants of rotting logs do occasionally make the dangerous journey from one log to another, and this shows that few environmental factors are absolute barriers to the dispersal of organisms and that they vary greatly in their effectiveness. Most habitats and microhabitats have only limited resources, and the organisms living in them must have mechanisms enabling them to find new habitats and resources when the old ones become exhausted. These mechanisms often take the form of seeds, resistant stages, or (as in the case of the insects of the rotting-log microhabitat) flying adults with a fairly high resistance to desiccation. There is good evidence that geographical barriers are not completely effective either. When organisms extend their distribution on a geographical scale, it is likely that they are taking advantage of temporary, seasonal, or permanent changes of climate or distribution of habitats that allow them to cross barriers normally closed to them. The British Isles, for instance, lie within the geographical range of about 220 species of birds, but a further 50 or 60 species visit the region as so-called 'accidentals' — these birds do not breed in Britain, but one or two individuals are seen by ornithologists every few years [4]. They come for a variety of reasons: some are blown off-course by winds during migration, others are forced in certain years to leave their normal ranges when numbers are especially high and food is scarce. Many of these accidentals have their true home in North America, such as the pectoral sandpiper (*Calidris melanotos*), a few of which are seen every year, but some come from eastern Asia, such as the olive-backed pipit (*Anthus hodgsoni*) or even from the South Atlantic, such as the black-browed albatross (*Diomedea melanophris*). It is possible, though not very likely, that a few of these chance travellers may in time establish themselves permanently in Europe, as did the collared dove (*Streptopelia decaocto*) which in a few decades has spread from Asia Minor and southern Asia across central Europe and into the British Isles and Scandinavia — perhaps the most dramatic change in distribution known in any vertebrate. This species is often common around the edges of towns and settlements, and seems to depend for food largely on the seeds of weed species common in farms and gardens. Several factors may have interacted to permit this extension of range of the collared dove. Increased human activity during the last century, involving extensive changes in the environment, has produced new habitats and food resources, and it is possible, too, that small changes in climate may have significantly favoured this species. It is, however, considered unlikely that the collared dove would have been able to take advantage of these changes without a change in its own genetic make-up, perhaps a physiological one permitting the species to tolerate a wider range of climatic conditions or to utilize a wider range of food substances.

Biogeographers commonly recognize three different types of pathway by which organisms may spread between one area and another. The first, easiest, pathway is called a *corridor*; such a pathway must include a wide variety of habitats, so that the majority of organisms found at either end of the corridor would find little difficulty in traversing it. The two ends would, therefore, come to be almost identical in their *biota* (i.e. the fauna plus the flora); for example, the great continent of Eurasia that links western Europe to China has acted as a corridor for the dispersal of animals and plants, at least until the recent climatic changes of the Ice Ages. Secondly, the interconnecting region may contain a more limited variety of habitats, so that only those organisms that can exist in these habitats will be able to disperse through it. Such a dispersal route is known as a *filter*; the exclusively tropical lowlands of Central America provide a good example. Finally, some areas are completely surrounded by totally different environments, so that it is extremely difficult for any organism to reach them. The most obvious example is the isolation of islands by wide stretches of sea, but the specially adapted biota of a high mountain peak, of a cave or of a large, deep lake is also extremely isolated from the nearest similar habitat from which colonists might originate. The chances of such a dispersal are therefore extremely low, and largely due to chance combinations of favourable circumstances, such as high winds or floating rafts of vegetation. Such a dispersal route is therefore known as a *sweepstakes route*. It differs from a filter in kind, not merely in degree, for the organisms that traverse a sweepstakes route are not normally able to spend their whole life histories *en route*. Instead, they are alike only in their adaptations to traversing the route, such as those aerial adaptations of spores, light seeds, insects, or birds that enable them to disperse from island to island. Such a biota is therefore not a representative sample of the ecologically integrated, balanced biotas of a normal mainland area, and is said to be *disharmonic*.

A discussion of some patterns of distribution shown by particular species of animals and plants will show how varied and complex these may be, and will help to emphasize the various scales or levels on which such patterns may be considered. In fact, the number of examples that we can choose is quite limited, because the distribution of only a very small number of species has been investigated in detail. Even amongst well-known species, chance finds in unusual places are constantly modifying known distribution patterns and requiring changes in the explanations that biologists give of these patterns.

Some existing patterns are continuous, the area occupied by the group consisting of a single region or of a number of regions which are closely adjacent to one another. These patterns can usually be explained by the distribution of present-day climatic and biological factors; the detailed distributions of several species of dragonfly, and of the plantains, provide good examples (below). Other existing patterns are discontinuous or *disjunct*, the areas occupied being widely separated and scattered over a particular continent, or over the whole world. The organisms which show such a pattern may, like the magnolias, be *evolutionary relicts*, the scattered survivors of a once-dominant and widespread group, now unable to compete with newer forms. Others, the *climatic relicts* or *habitat relicts* appear to have been greatly affected by past changes in climate or sea level. Finally, as will be shown in Chapters 7 and 8, the disjunct patterns of some living groups, and of many extinct groups, have resulted from the physical splitting of a once-continuous area of distribution by the process of continental drift.

The cosmopolitan plantain

The broad-leaved plantain, *Plantago major*, has a distribution that could be described as cosmopolitan, because it is found on all the continents except Antarctica [5] (Fig. 3.4). It is typically a species of grassland habitats, and has a rosette of broad leaves pressed close to the soil surface, from which flower-bearing stems rise. Its distribution has not been as thoroughly studied as that of the dragonflies discussed below, but it appears to be ubiquitous except for the higher northern latitudes and the deserts of Africa and the Middle East. It has even spread through the East Indies to Australia, New Zealand, and many Pacific islands. On a regional scale, the distribution of *P. major* within the British Isles is well known; the species is found almost everywhere and appears to be quite unaffected by variations in climate or in soil conditions — it is found in grassland on both very acid and very alkaline soils. An ecologist would describe this species as an *ecologically tolerant* or *eurytopic* one, because its habitat preferences are so broad. But this alone is not enough to account for its wide distribution; it must have a highly efficient dispersal mechanism. This is provided by its seeds, some of which may be eaten by birds and subsequently dropped in a new habitat. The seeds are resistant to environmental conditions, including those of animals' digestive systems, and at least a proportion of the

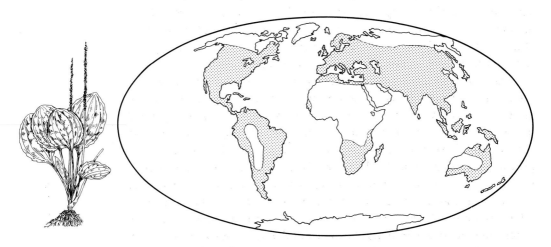

Fig. 3.4 The world distribution of a cosmopolitan species, *Plantago major*, the broad-leaved plantain.

seeds eaten by animals can germinate after passing through the gut. The seeds also have a coat of mucilage surrounding them, which renders them adhesive. They stick very easily to feathers and fur and may be transported from one place to another in this manner. It is also probable that man has played a part in the dispersal of this plant, because it is quite likely to be mixed with grasses cut for hay and subsequently carried long distances over land or sea. This seems to be its most likely method of transport across the Pacific Ocean to South America. Seeds may also adhere to the tracks and tyres of vehicles and be carried considerable distances overland in this way.

Since *P. major* is not in fact found everywhere, what are the factors limiting its distribution? As with many other organisms, the full answer is not yet known. The plant does not extend far into northern regions, but it is found at quite high altitudes elsewhere, and cold is therefore probably not a limiting factor. It is not found in any really arid areas; its broad leaves offer a large surface for evaporation of water and it may become desiccated in dry climates. Local distribution within its grassland habitat provides some other clues. Areas of high abundance occur around gateways and footpaths, and in a few other areas where grazing animals might collect. This is because the plantain is most common in places where there is intensive grazing by cattle, and trampling by these and other animals, including man. Because of its flattened form it can withstand these pressures better than most grassland plants. *P. major* probably grows best in such places because it needs full sunlight for efficient photosynthesis. In other situations it is usually shaded from sunlight by taller plants and is unable to grow well. The species also tends to be an early colonizer of disturbed, bare soil but is eliminated from such places when other species grow tall. Also its seeds need to be unshaded in order to germinate efficiently. It is this germination and establishment phase that is most sensitive both to climatic stress and to competition. Colonization of the far north by this species is probably prevented by its failure to germinate and establish healthy adult plants in tundra conditions.

Patterns of dragonflies

One group of species whose distributions are quite well known, at least in western Europe, are the Odonata, dragonflies and damselflies [6]. The common blue damselfly, *Enallagma cyathigerum*, is possibly one of the most abundant and widely distributed dragonfly species (Fig. 3.5). The adults are on the wing in mid-summer, around bodies of fresh water. The female lays eggs in vegetation below the surface of the water and the larvae hatch in a week or two. These live on the bottom of the pond, stream or lake and feed on small crustaceans and insect larvae until

Fig. 3.5 The distribution on a world scale of three species of dragonfly. (a) *Enallagma cyathigerum*, (b) *Sympetrum sanguineum*, and (c) *Anax imperator*.

they reach a size of 17–18 mm, which may take from two to four years, depending on the quality and quantity of food available. In the May or June after reaching full size, the larvae climb up the stems of *emergent vegetation* (plants rooted in the mud with stem and leaves emerging from the water surface), cast their larval skins, and emerge as winged adults. *E. cyathigerum* is found in a wide range of freshwater habitats, including both still and moving water, though it is perhaps least common in fast-moving water, or in places where silt is being deposited. Probably the ideal habitat for this species is a fairly large body of still water with plenty of floating vegetation. *Enallagma cyathigerum* is found in both acid and alkaline waters, and often occurs in brackish pools on salt marshes. As a result, clearly, this species is eurytopic, like *Plantago*.

The geographical distribution of this damselfly is also very wide. In Europe the distribution lies mostly between 45°N and the Arctic Circle, although it includes some of the wetter parts of Spain, and is rather scattered in northern Scandinavia; it is not found in Greenland or Iceland. The species is found in a few places in North Africa, in Asia Minor, and around the Caspian Sea. Large areas of Asia south of the Arctic Circle also fall within its range. In North America it is found everywhere north of about 35–40°N to the Arctic Circle, except Labrador and Baffin Island. Populations also occur in the ideal habitats provided by the Everglades of Florida, which mark the species' furthest southward expansion. The broad geographical distribution of this species is almost certainly due to its ecological tolerance and ability to make use of a wide range of habitats in very different climates. This type of distribution pattern, a belt around the Northern Hemisphere, is shown by many species of animals and plants, and is termed *circumboreal* — 'around the northern regions'. The frequency of this pattern in very different organisms suggests that the two northern land masses may once have been joined, enabling certain species to spread right around the hemisphere.

As *E. cyathigerum* is so successful, one might ask why it has not spread further southward. One reason may be the relative scarcity of watery habitats in the subtropical regions immediately to the south of its present range — the arid areas of Central America, North Africa, and central Asia. The species is perhaps not robust enough for the long migrations that would be needed to reach suitable habitats in the Southern Hemisphere (there are very few wind belts that might assist such a migration). Another possibility is that there are other species already occupying all the habitats that *E. cyathigerum* could colonize further south. These species may be better adapted to the physical conditions of their habitats than *E. cyathigerum*, and could therefore compete successfully with it for the available food resources; this might exclude the species from these areas. In fact in the Southern Hemisphere there are many species of the genus *Enallagma* and the closely related genus *Ischnura* that might be expected to have similar habitat and food requirements to *E. cyathigerum*; there are at least eight species of *Enallagma* in South Africa alone. Any of these explanations, or a combination of them, would explain why the common blue damselfly is confined to the high latitudes of the Northern Hemisphere.

The distribution of *E. cyathigerum* may be contrasted with that of the beautiful dragonfly *Sympetrum sanguineum*, sometimes called the ruddy sympetrum (Fig. 3.5b). This species has a limited distribution in western Europe, parts of Spain, a few places in Asia Minor and North Africa, and around the Caspian Sea; it is not found in eastern Asia or North America. The reason for the limited distribution of this dragonfly is almost certainly that the larva has very precise habitat requirements.

The larva is found in ditches and ponds with still waters, but only where certain emergent plants — the great reedmace (*Typha latifolia*) and horsetails (*Equisetum*) — are growing. Why the larva should have these precise requirements is not clear, because it is certainly not a herbivore, and feeds on insect larvae and crustaceans, but so far the larvae have never been found away from the roots of these plants. *Sympetrum sanguineum* could therefore be described as a *stenotopic* species — one with very limited ecological tolerance. The fact that it can colonize only a very few habitats must certainly have limited its distribution, but other, unknown, factors are also at work, for the species is often absent from waters in which reedmace or horsetails are present.

The northern distribution of these two species may be contrasted with that of the emperor dragonfly, *Anax imperator* [7] (Fig. 3.5c). The adults of this species are 8—10 cm long, and the larva is found typically in large ponds and lakes and in slow-moving canals and streams; it is a voracious predator and can eat animals as big as fish larvae. The distribution covers a band of Europe between about 50°N and 40°N but, unlike the other two species, it is well distributed on the North African coast and the Nile Valley and stretches across Asia Minor to north-west India. It even spreads across the Sahara Desert down into Central Africa, where there are suitable habitats such as Lake Chad and the lakes of East Africa, and it is found in most parts of South Africa except the Kalahari Desert. (It is possible that the South African population may belong to a separate subspecies from the European forms.)

So the distribution of *A. imperator* is confined to the Old World and does not extend far into Asia. It appears to be basically a Mediterranean and subtropical species whose good powers of flight and fairly broad ecological tolerance have enabled it to cross the unfavourable areas of North Africa to new habitats in southern Africa. No doubt favourable habitats for *A. imperator* do occur in the other land masses (although there may be potential competitors there, of course), but the dragonfly cannot now reach them because the land connections lie to the north, where its distribution seems limited.

In most tropical dragonfly species the larvae emerge from the water and metamorphose to the adult at night; they are very vulnerable to predators at the time of emergence, and darkness probably affords some protection from birds. But the process of metamorphosis is inhibited by cold temperatures, and in northern Europe many species are compelled by low night temperatures to undergo at least part of their emergence in daylight, when birds eat large numbers of them. This probably puts a northern limit of the distribution of many species of dragonfly, including *A. imperator*, which would explain why this species has not been able to invade the Americas or eastern Asia.

Magnolias — evolutionary relicts

The magnolias (family Magnoliaceae, genus *Magnolia*) have a very interesting modern distribution, as shown in Fig. 3.6. Of the 80 or so species of *Magnolia*, the majority are found in south-east Asia and the remainder, about 26 species, in the Americas — ranging from Ontario in the north, through Mexico, down into the northern regions of South America [8]. Their distribution is clearly disjunct, being separated into two main centres in this case. Unlike the palms, we cannot explain this distribution simply in terms of the climatic sensitivities of the plants concerned, for the magnolias are reasonably hardy in comparison; they can be cultivated well into the north of the temperate area. Nor would climatic constraints explain why they are not found in intermediate tropical and subtropical regions, as are the palms.

To understand the distribution of the magnolias, we need to look at their evolutionary history. Fossils of magnolia-like leaves, flowers and pollen grains are known from Mesozoic times — the age of the dinosaurs. Indeed, the magnolia family is regarded by botanists as one of the most primitive families of flowering plants. Its showy flowers were attractive to the rapidly evolving insects and together they coevolved into a most successful team in which the insect visited the flowers for food and, in doing so, ensured the passage of pollen from one plant to another, thus taking the chance and the waste out of the highly risky pollination process. The magnolias spread and must have formed a fairly continuous belt around the tropical, subtropical and temperate parts of the world, for their fossil remains have been found through Europe and even in Greenland. For perhaps as long as 70 million years, the magnolias remained widespread, right up to the last two million years, during

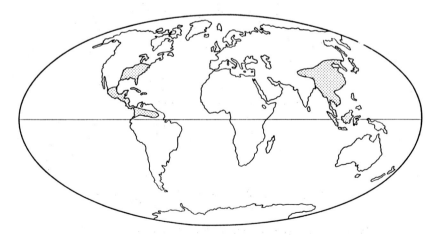

Fig. 3.6 World distribution map of the magnolias, illustrating a disjunct distribution.

which they have been lost from such intermediate geographical areas as Europe.

Being small, slow-growing shrubs and trees, they were not strong competitors for the more robust and fast-growing tree species and, when the climatic fluctuations of the last two million years began to disturb their stable woodland environment, they succumbed to the pressures and soon became extinct across much of their former range. Only in two parts of the world have they managed to escape and survive, as evolutionary relicts.

It is interesting that another genus of the magnolia family, the tulip tree, *Liriodendron*, has a very similar distribution to that of the *Magnolia* genus and in all probability shares a similar fossil history. But, in the case of the tulip trees, only two species have survived, *L. tulipifera* being a successful component of the deciduous, temperate forests of North America, and *L. chinense* surviving only in a restricted area of south-east Asia (Fig. 3.7).

Climatic relicts

Many other species, which in the past were widely distributed, were affected by climatic changes and survive now only in a few 'islands' of favourable climate. Such species are called *climatic relicts* — they are not necessarily species with long evolutionary histories, since many major climatic changes have occurred quite recently. The Northern Hemisphere has an interesting group of *glacial relict* species whose distributions have been modified by the northward retreat of the great ice sheets that extended as far south as the Great Lakes in North America, and to Germany in Europe, during the Pleistocene Ice Ages

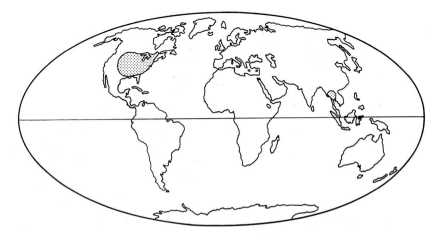

Fig. 3.7 World distribution of the tulip trees (*Liriodendron* species). Only two species now survive, in widely separated localities, though it was once a widespread genus.

(the last glaciers retreated from these temperate areas about 10 000 years ago). Many species that were adapted to cold conditions at that time had distributions to the south of the ice sheets almost as far as the Mediterranean. Now that these areas are much warmer, these species survive there only in the coldest places, usually at high altitudes in mountain ranges, and the greater part of their distribution lies far to the north in Scandinavia, Scotland, or Iceland. In some cases, species even appear to have become extinct in northern regions and are represented now only by glacial relict populations in the south.

An interesting glacial relict is the springtail *Tetracanthella arctica* (Insecta, Collembola) [9]. This dark blue insect, only about 1.5 mm long, lives in the surface layers of the soil and in clumps of moss and lichens, where it feeds on dead plant tissues and fungi. It is quite common in the soils of Iceland and Spitzbergen, and has also been found further west in Greenland and in a few places in Arctic Canada. Outside these truly Arctic regions it is known to occur in only two regions; in the Pyrenean Mountains between France and Spain, and in the Tatra Mountains on the borders of Poland and Czechoslovakia (with isolated finds in the nearby Carpathian Mountains) (Fig. 3.8). In these mountain ranges the species is found at altitudes of around 2000 m in arctic and subarctic conditions. It is hard to imagine that the species can have colonized these two areas from its main centre further north, because it has very poor powers of distribution (it is quickly killed by low humidity or high temperatures) and is not likely to have been transported there accidentally by man. The likely explanation of the two southern populations is that they are remnants of a much wider distribution in Europe in the Ice

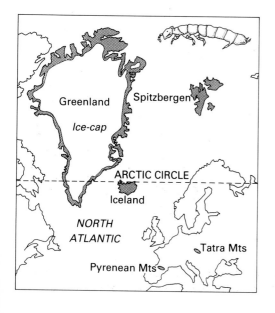

Fig. 3.8 The springtail *Tetracanthella arctica*, and a map of its distribution. It is found mostly in northern regions, but populations exist in the Pyrenees and in mountains in central Europe. These populations were isolated at these cold, high altitudes when the ice sheet retreated northwards at the end of the Ice Ages.

Ages. But it is surprising that *T. arctica* has not been found at high altitudes in the Alps, despite careful searching by entomologists. Perhaps it has simply not yet been noticed, or perhaps it used to occur there but has since died out. One interesting feature of this species is that whereas animals from the Arctic and the Tatras have eight small eyelets (*ocelli*) on either side of the head, specimens from the Pyrenees have only six. This suggests that the Pyrenean forms have undergone some evolutionary changes since the end of the Ice Ages while they have been isolated from the rest of the species, and perhaps they should be classified as a separate subspecies.

A plant example of a glacial relict (see Fig. 3.9) is the Norwegian mugwort (*Artemisia norvegica*), a small alpine plant now restricted to Norway, the Ural Mountains and two isolated localities in Scotland. During the last glaciation and immediately following it, the plant was widespread, but it became restricted in distribution as forest spread.

There are probably several hundred species of both animals and plants in Eurasia that are glacial relicts of this sort, and they include many species that, in contrast to the springtail, have quite good powers of distribution. One such species is the mountain or varying hare, *Lepus timidus*, a seasonally variable species (its fur is white in the winter and bluish for the rest of the year), which is closely related to the more common brown hare, *L. europaeus*. The varying hare has a circumboreal distribution, including Scandinavia, Siberia, northern Japan and northern Canada (although the North American forms are thought by some zoologists to form a separate species, *L. americanus*). The southernmost

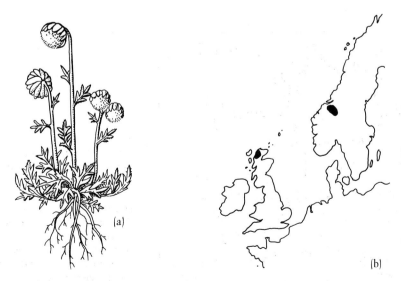

Fig. 3.9 The Norwegian mugwort, *Artemisia norvegica*: (a) the plant, (b) distribution map showing its restricted range in only two mountainous areas of Europe.

part of the main distribution is in Ireland and the southern Pennines of England, but there is a glacial relict population living in the Alps that differs in no important features from those in the more northerly regions. There is, however, an interesting complication — *L. timidus* is found all over Ireland, thriving in a climate that is no colder than that of many parts of continental western Europe. There seems to be no climatic reason why this hare should not have a wider distribution, but it is probably excluded from many areas by its inability to compete with its close relative, the brown hare, for food resources and breeding sites. Relict populations of the varying hare survive in the Alps because, of the two species, it is the better adapted to cold conditions [10].

One very remarkable example of a glacial relict is the dung beetle species *Aphodius holdereri* (Fig. 3.10). This beetle is now restricted to the high Tibetan plateau (3000–5000 m), having its southern limit at the northern slopes of the Himalayas. In 1973 G.R. Coope, of Birmingham University in England, found the remains of at least 150 individuals of this species in a peaty deposit from a gravel pit at Dorchester-on-Thames in southern England [11]. The deposit dated from the middle of the last glaciation, and subsequently 14 sites have yielded remains of this species in Britain, all dated between 25 000 and 40 000 years ago. Evidently *A. holdereri* was then a widespread species, but climatic changes have severely restricted the availability of suitable habitats for its survival. Only the remote Tibetan mountains now provide *Aphodius* with the extreme climatic conditions necessary for its survival.

The strawberry tree in Europe is a good example of what may be termed a post-glacial relict (Fig. 3.11). The strawberry trees of North America have a fairly continuous distribution, but the western European species *Arbutus unedo* is disjunct, having its main centre of distribution in the Mediterranean region but with outliers in western France and western Ireland.

The Ice Age closed with a sudden warming of the climate, and the

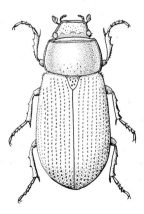

Fig. 3.10 *Aphodius holdereri*, a dung beetle now found only in the high plateau of Tibet.

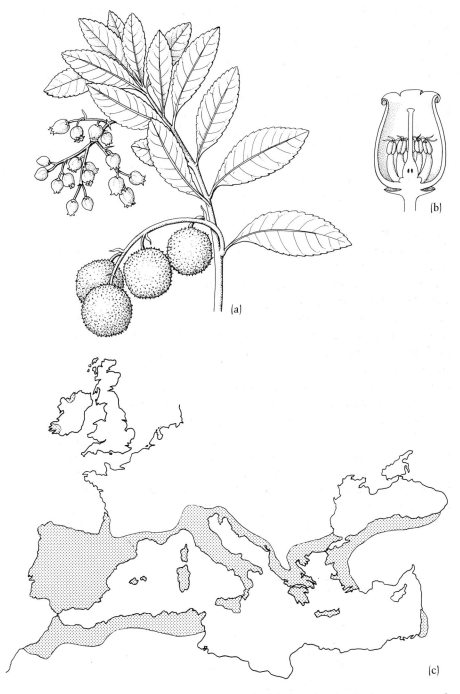

Fig. 3.11 The strawberry tree, *Arbutus unedo*: (a) plant showing leathery leaves, and swollen fruit which are red in colour; (b) cross section of a flower; (c) map of European distribution, showing relict population in Ireland.

glaciers retreated northwards; behind them came the plant and animal species that had been driven south during glacial times. Warmth-loving animals, particularly insects, were able to move northward rapidly. Plants were slower in their response, because their rate of spread is slower. Seeds were carried northward, germinated, grew, and finally flowered and sent out more seeds to populate the bare northlands. As this migration continued, melting glaciers produced vast quantities of water that poured into the seas, and the ocean levels rose. Some of the early colonizers reached areas by land connections that were later severed by rising sea levels.

The maritime fringe of western Europe must have provided a particularly favourable migration route for southern species during the period following the retreat of the glaciers. Many warmth-loving plants and animals from the Mediterranean region, such as the strawberry tree, moved northward along this coast and penetrated at least as far as the south-west of Ireland, before the English Channel and the Irish Sea had risen to form physical barriers to such movement. The nearness of the sea, together with the influence of the warm Gulf Stream, gives western Ireland a climate that is wet, mild, and frost-free, and this has allowed the survival of certain Mediterranean plants that are scarce or absent in the rest of the British Isles.

Like many Mediterranean trees and shrubs, the strawberry tree is *sclerophyllous*, which means it has hard, leathery leaves (see Fig. 3.11). This is a plant adaptation often associated with arid climates. Flowering in many plant species is triggered by a response to a particular daylength — this is called *photoperiodism*. *Arbutus* flowers in autumn, as the length of night is increasing, and this is an adaptation which is again associated with Mediterranean conditions, since at this season the summer drought gives way to a warm, damp period. The flowers, which are cream-coloured, conspicuous, and bell-shaped, have nectaries that attract insects, and in Mediterranean areas they are pollinated by long-tongued insects such as bees, which are plentiful in late autumn. In Ireland, however, insects are scarce in the autumn and pollination is therefore much less certain. Thus the strawberry tree reached Ireland soon after the retreat of the glaciers and has since been isolated there as a result of rising oceans. Although the climate has steadily grown colder since its first colonization, *Arbutus* has so far managed to hold its own and survive in this outpost of its range.

Another example of a disjunction that has taken place in post-glacial times is the gorilla (*Gorilla gorilla*). In West Africa it is found in an area of lowland tropical rain forest, but in the east it is not limited to lowland forest but is found in mountains also (Fig. 3.12). Some zoologists regard the two populations as separate subspecies. The two areas are now separated by savanna grassland across which the gorillas cannot

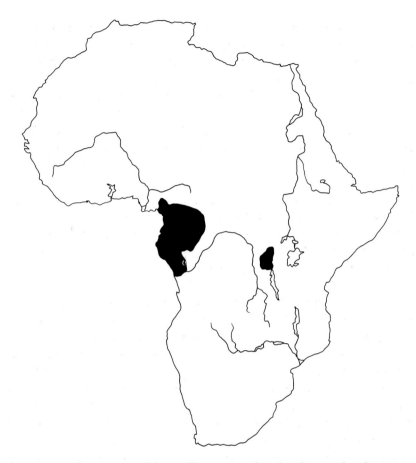

Fig. 3.12 Distribution map of the gorilla, a mammal with a disjunct distribution.

migrate, but they were probably in contact with one another in relatively recent (Pleistocene) times (see p. 253). This pattern of disjunction is reflected in the distributions of many African plants and animals [12].

Endemic organisms

Because each new species of organism evolves in one particular, restricted area, its distribution will be limited by the barriers that surround its area of origin. Each such area will, therefore, contain organisms that are found there and nowhere else; these organisms are said to be *endemic* to that area. As time goes by, more and more organisms will evolve within the area, and the percentage of its biota that is endemic is therefore a good guide to the length of time for which an area has been isolated.

As these organisms continue to evolve they will also become progressively different from their relatives in other areas. Taxonomists are likely to recognize this by raising the taxonomic rank of the organisms concerned. So, for example, after two million years the biota of an isolated area might contain only a few endemic species. After 10 million years the descendants of these species might be so unlike their nearest relatives in other areas that they might be placed in one or more endemic genera. After 35 million years these genera might appear to be sufficiently different from their nearest relatives as to be placed in a different family, and so on. (The absolute times involved would, of course, vary and depend upon the rate of evolution of the group in question.) So, the longer an area has been isolated, the higher the taxonomic rank of its endemic organisms is likely to be, and vice-versa.

Figure 3.13 shows the proportion of the montane flora in various European mountain ranges that are endemic to their particular area. It is evident that the more northerly of the mountain ranges shown have a lower proportion of their flora that is endemic whereas the southern, Mediterranean mountains, have higher proportions [13].

The montane plants, like the glacial relicts described above, are now limited in range because of the increasing warmth of the last 10 000 years. The northern mountains may be poorer in endemics simply because local glaciation there was more severe and some of the species surviving further south became extinct. On the other hand, the richness of the southern mountains could be explained by the fact that the geographical barriers between the northern montane blocks are less severe (less distance, no sea barriers) and hence migration and sharing of mountain floras is more likely than in the south, where barriers are considerable.

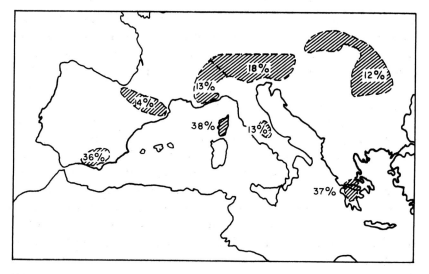

Fig. 3.13 The percentage of endemic plants in the floras of the mountain ranges of southern Europe. (From Favarger [13].)

In general there are two major factors influencing the degree of endemism in an area; these are isolation and stability. Thus, isolated islands and mountains are often rich in endemics. Long-term climatic stability is rather rare, but there is evidence that some parts of the earth have been more stable than others. For example, the Cuatro Ciénegas basin in Mexico appears to have retained a very stable vegetation over the last 40000 years, judging from the evidence of pollen from lake cores collected in the area (see Chapter 10). This basin is also rich in endemic organisms [14].

California is another region which is very rich in endemics, and this is again a somewhat isolated area, with high mountains and deserts separating it from most of the rest of America. Once more it is possible to relate some of the endemics to long survival in isolated areas. On the other hand, many plant species in California are undergoing rapid evolution, and new types are arising in such plant genera as *Aquilegia* and *Clarkia*. These so-called *neoendemics* (as opposed to the ancient *palaeoendemics*) may be restricted simply because they have not had time to expand their ranges into other areas. G.L. Stebbins and J. Major proposed that the Californian palaeoendemics were restricted mainly to either wet or arid areas, whereas the neoendemics were found mainly in less predictable habitats, and subsequent work has supported this idea [15].

Physical limitations

The geographical range of a species is not always determined by the presence of topographic barriers preventing its further spread. Often a species' distribution is limited by a particular factor in the environment that influences its ability to survive or reproduce adequately. These limiting factors in the environment include *physical factors* such as temperature, light, wetness, and dryness, as well as *biotic factors* such as competition, predation, or the presence or absence of suitable food. In the remainder of this chapter the ways in which such factors influence organisms will be described in more detail.

First, though, the meaning of the term 'limiting factor' must be understood. Anything that tends to make it more difficult for a species to live, grow, or reproduce in its environment is a limiting factor for the species in that environment. To be limiting, such a factor need not necessarily be lethal for a species; it may simply make the working of its physiology or behaviour less efficient, so that it is less able to reproduce or to compete with other species for food or living space. For instance, we suggested earlier that a northern limit may be set to the distribution of certain dragonflies by low night-time temperatures. In the more southerly parts of northern regions, at least, temperatures are not so low that they kill dragonflies *directly*, but they are low enough at night to force the insects to metamorphose during the day, when they are more vulnerable to predatory birds. In this case, then, the limiting factor of temperature does not operate directly but is connected with a biotic environmental factor, that of predation. Many other limiting factors act in a similar way.

Environmental gradients

Many physical and biotic factors affect any species of organism, but each can be considered as forming a *gradient*. For example, the physical factor of temperature affects species over a range from low temperatures at one extreme to high temperatures at the other, and this constitutes a temperature gradient. These gradients exist in all environments and affect all the species in each environment. As seen earlier, different species vary in their tolerance of environmental factors, being either *eurytopic* (ecologically tolerant) or *stenotopic* (ecologically intolerant), but each species can function efficiently over only a more or less limited part of each gradient. Within this *range of optimum* the species can survive and maintain a large population; beyond it, toward both the low and the high ends of the gradient, the species suffers increasing physiological stress — it may stay alive, but because it cannot function efficiently it can maintain only low populations. These areas of the gradient are bordered by the upper and lower limits of tolerance of the species to the environmental factor. Beyond these limits the species cannot survive because conditions are too extreme; individuals may live there for short periods but will either die or pass quickly through to a more favourable area (Fig. 3.14). A species may not achieve its full potential distribution in the field because of competitive interactions with other organisms. When under conditions of physiological stress, a species easily succumbs to such competition.

The grey hair grass (*Corynephorus canescens*) is widespread in central and southern Europe and reaches its northern limit in the British Isles and southern Scandinavia (Fig. 3.15). J.K. Marshall has examined the factors that may be responsible for maintaining its northern limit, and he found that both flowering and germination were affected by low temperature [17]. The grass has a short life span (about two to six years), so it relies upon seed production to maintain its population. Any factor interfering with flowering or with germination could therefore limit its success in competitive situations. At its northern limit, low summer temperature delays its flowering, with the result that the season is already well advanced when the seeds are shed. Seed germination is slowed down at temperatures below 15°C and seeds sown experimentally after October had a very poor survival rate. This may explain why its northern limit in Europe so closely matches the 15°C July mean isotherm.

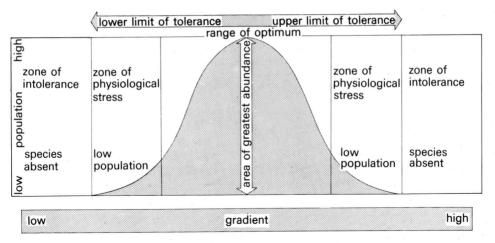

Fig. 3.14 Graphical model of the population abundance maintained by a species of animal or plant along a gradient of a physical factor in its environment. From Kendeigh [16].

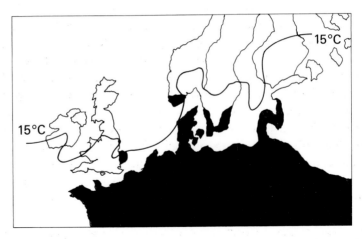

Fig. 3.15 Distribution of the grey hair grass (*Corynephorus canescens*) in northern Europe (shaded) and its relationship to the 15°C July mean isotherm. From Marshall [17].

Even mobile animals, like birds, may have their distributions closely linked to temperature, as in the case of the eastern phoebe (*Sayornis phoebe*), a migratory bird of eastern and central North America. Analysing data collected by ornithologists of the National Audubon Society during the Christmas period, ecologist Terry Root has been able to check the winter distribution of this bird against the climatic conditions [18]. She found that the wintering population of the eastern phoebe was confined to that part of the United States in which the mean minimum January temperature exceeded −4°C. The very close correspondence of the bird's

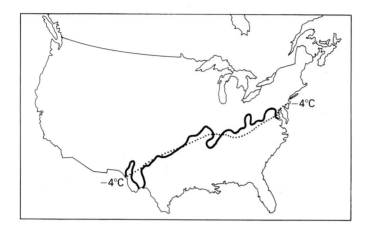

Fig. 3.16 The northern boundary of the distribution (solid line) of the eastern phoebe in North America in December/January, compared with the −4°C January minimum isotherm. From Root [18].

winter range to this isotherm (see Fig. 3.16) probably relates to the energy balance of the birds. Warm-blooded animals, like birds, use up large quantities of energy in maintaining their high blood temperature and in cold conditions they can lose a great deal of energy in this way and therefore have to eat more. Terry Root found that birds in general do not occupy regions where low temperature forces them to raise their resting metabolic rate (i.e. their energy consumption) by a factor of more than 2·5. In the case of the eastern phoebe this critical point is reached when the temperature falls below 4°C, so the bird fails to occupy colder regions. Other birds have different temperature limits because they have different efficiencies in their heat generation and conservation, but they still seem to draw the line at raising their resting metabolism by more than 2·5 times.

Many plants have their seeds adapted to a specific temperature for germination, and this often relates to conditions prevailing when germination is most appropriate for the species. P.A. Thompson of Kew Gardens, England, has devised a piece of apparatus for examining the effect of temperature on germination [19]. It consists of a metal bar, one end of which is maintained at −40°C and the other at 3°C; between is a gradient of temperatures. Groups of seeds of the species to be examined are placed along the bar and kept moist, and a record is kept of the number of days required for 50 per cent of the seeds within each group to germinate. The results are expressed on graphs, and the lowest point on the U-shaped curves shows the optimum temperature for germination.

In Fig. 3.17 the germination responses of three members of the catchfly family are shown, together with their geographical ranges. The catchfly (*Silene secundiflora*) is a Mediterranean species, so the optimum

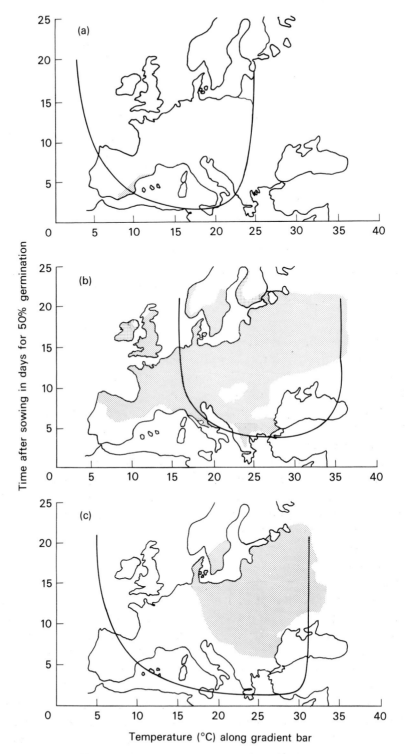

Fig. 3.17 Distribution maps of three members of the plant family Caryophyllaceae, together with their germination responses to temperature: (a) *Silene secundiflora*, (b) *Lychnis flos-cuculi* and (c) *Silene viscosa*. From Thompson [19].

time for germination is the autumn, when the hot, dry summer is over
and the cool, moist winter is about to begin. Its optimum germination
occurs at about 17°C. The ragged robin (*Lychnis flos-cuculi*) occurs
throughout temperate Europe and here the cold winter is the least
favourable for growth, hence there are advantages to be gained by germi-
nating in the spring. Optimum germination occurs at about 27°C. The
third species, the sticky catchfly (*Silene viscosa*), is an eastern European
steppe species. The invasion of open grassland is an opportunistic busi-
ness; each chance that offers itself must be taken, so any temperature
limitation is likely to be an unacceptable restriction on a plant in its
struggle for space. Wide tolerance of temperature is thus an advantage
and *S. viscosa* seeds germinate well over the range 11−31°C.

In most plants the first product of photosynthesis is a sugar containing
three carbon atoms; these plants are known as *C_3 plants*. In some plant
species, however, there is a supplementary mechanism at work in which
carbon dioxide is temporarily fixed into a four-carbon compound, later
to be fed into the conventional fixation process in specialized cells
around the bundles of conducting tissue in the leaf [20]. These *C_4 plants*
occur in a number of different flowering plant families, mostly from
tropical and subtropical regions. Some important tropical crop species,
like sugar cane, are C_4 plants.

For a variety of biochemical reasons the C_4 mechanism is most
advantageous under conditions of high light intensity and high temperature, whereas it may be disadvantageous at low light and low temperature
[21]. J.R. Ehleringer has calculated that, at a latitude of about 45° in an
area like the great plains of North America, the relative advantages and
disadvantages of each system are roughly in balance (Fig. 3.18) [22]. If
one examines the proportion of C_3 and C_4 species among the grasses in
different sites in North America, then one finds that the C_4 system
occurs in more than 50 per cent of the grasses in most areas south of
40°N and in less than 50 per cent north of that latitude (Fig. 3.19) [23].
So competitive interaction between species of grass has led to the
selection of the photosynthetic mechanism most appropriate to the
needs of any given locality. Those C_4 species found north of the critical
line are often associated with particular circumstances that favour
them; for example, they may have their maximum growth rate in late
summer when temperatures are highest, whilst the C_3 species grow best
in the cooler conditions of spring and early summer [24].

Although temperature is one of the most important environmental
factors, because of its effect on the metabolic rate of organisms, many
other physical factors in the environment are limiting ones. A whole
family of factors is related to the amount of water present in the
environment. Aquatic organisms obviously require water as the basic
medium of their existence, but most terrestrial animals and plants, too,

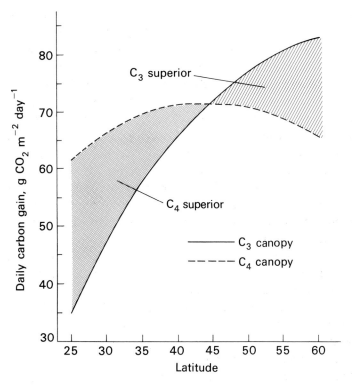

Fig. 3.18 Predicted levels of photosynthesis for C_3 and C_4 species over a range of latitudes in the Great Plains during July. The C_4 advantage is lost in latitudes greater than 45°N. From Ehleringer [22].

are limited by the wetness or dryness of the habitat, and often also by the humidity of the atmosphere, which in turn affects its 'drying power' (or, more precisely, the rate of evaporation of water from the ground and from animals and plants). Light is of fundamental importance because it provides the energy that green plants fix into carbohydrates during photosynthesis, thus obtaining energy for themselves (and ultimately for all other organisms).

Light in its daily and seasonal fluctuation also regulates the activities of many animals. The concentrations of oxygen and carbon dioxide in the water or air surrounding organisms are also important. Oxygen is essential to most animals and plants for the release of energy from food by respiration, and carbon dioxide is vital because it is used as the raw material in the photosynthesis of carbohydrates by plants. Many other chemical factors of the environment are of importance, particularly soil chemistry where plants are concerned. Pressure is important to aquatic organisms; deep-sea animals are specially adapted to live at high pressures, but the tissues of species living in more shallow waters would be easily damaged by such pressures.

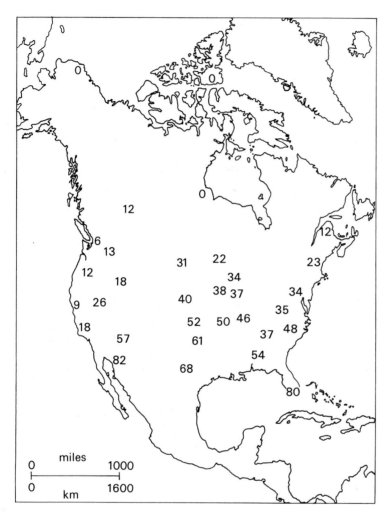

Fig. 3.19 Proportion of C$_4$ species in the grass flora of various parts of North America. From Teeri and Stowe [23].

In marine environments, variation in the salinity of the water affects many organisms, because many marine organisms have body fluids with much the same salt concentration as sea water (about 35 parts per thousand), in which their body tissues are adapted to function efficiently. If they become immersed in a less saline medium (in estuaries, for instance), water moves into their tissues due to the physical process called *osmosis*, by which water passes from a dilute solution of a salt to a concentrated one. If the organisms cannot control the passage of water into their bodies, the body fluids are flooded and their tissues can no longer function. This problem of salinity is an important factor in preventing marine organisms from invading rivers, or freshwater ones from invading the sea and spreading across oceans to other continents (Fig. 3.20).

Interaction of factors

The environment of any species consists of an extremely complicated series of interacting gradients of all the factors, biotic as well as physical, and these influence its distribution and abundance. Populations of the species can live only in those areas where favourable parts of the environmental gradients that affect it overlap. Factors that fall outside this favourable region are limiting ones for the species in that environment.

Some of the interactions between the various factors in an organism's environment may be very complex and difficult for the ecologist to interpret, or for the experimentalist to investigate. This is because a series of interacting factors may have more extreme effects on the

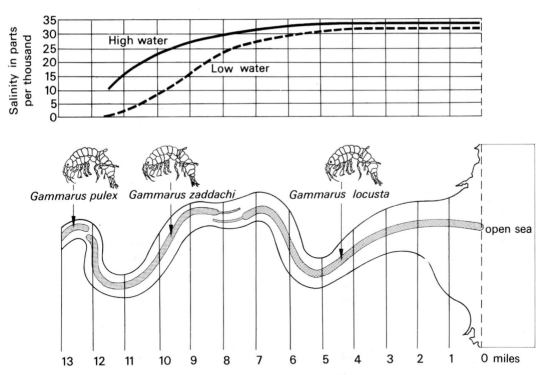

Fig. 3.20 The distribution along a river of three closely related species of amphipod (Crustacea), relative to the concentration of salt in the water. *Gammarus locusta* is an estuarine species and is found in regions where salt concentration does not fall below about 25 parts per thousand. *G. zaddachi* is a species with a moderate tolerance of salt water and is found along a stretch of water between 8 and 12 miles (11–19 km) from the river mouth where salt concentrations average 10–20 parts per thousand. *G. pulex* is a true freshwater species and does not occur at all in parts of the river showing any influence of the tide or salt water [25].

behaviour and physiology of a species than any factor alone. To take a simple example, temperature and water interact strongly on organisms, because both high and low temperatures reduce the amount of water in an environment, high temperatures causing evaporation and low ones causing freezing, but it may be very hard to discover if an organism is being affected by heat or cold or by lack of water. Similarly, light energy in the form of sunlight exerts a great influence on organisms because of its importance in photosynthesis and in vision, but it also has a heating effect on the atmosphere and on surfaces, and therefore raises temperatures. In natural situations it is often almost impossible to tell which of many possible limiting factors is mainly responsible for the distribution of a particular species.

An interesting example of the complexity of interaction between environmental factors was studied by the American ecologist M.R. Warburg, in his work on two species of woodlice (sowbugs or slaters, Crustacea, Isopoda) living in rather dry habitats in southern Arizona [26]. One species, *Armadillidium vulgare*, is found mostly in grasslands and scrubby woodland and is also widely distributed in similar habitats elsewhere in North America and in Europe. The other, *Venezilla arizonicus*, is a rather rare species, confined to the south-western United States and found in very arid country, with stony soil and a sparse vegetation of cactus and acacia. Warburg investigated the reactions of these two species to three environmental factors: temperature, atmospheric humidity and light. His experimental techniques involved the use of a simple apparatus, the choice chamber or *preferendum apparatus*, in which animals may be placed in a controlled gradient of an environmental factor. The behaviour of the animals, particularly the direction in which they move and their speed, can then be used to suggest which part of the gradient they find most satisfactory; this is termed the preferendum of these particular animals in this particular gradient. Warburg's method for testing the interactions of light, temperature and humidity on the woodlice was the classic scientific approach of isolating the effects of each factor separately, and then testing them two or three at a time in all possible combinations. For instance, he might set up a gradient of temperature between hot and cold and test the reactions of the animals to this, either with the whole gradient at a low humidity (dry) or the whole gradient at a high humidity (wet), or with the hot end of the gradient dry and the cold wet, or with the cold end dry and the hot end wet. He might then test the effect of light on these four situations by exposing each in turn to constant illumination, constant darkness, or one end of the gradient in darkness and the other in light. Such work is extremely time-consuming and requires great patience.

Warburg found that, in general, *A. vulgare* prefers low temperatures (around 10–15°C), high humidities (above 70 per cent relative humidity, i.e. the air is 70 per cent saturated with water vapour) and is rather weakly attracted to light. This accords well with what is known of the species' habitat and habits — it lives in fairly humid, cool places and is active during the day. *Venezilla arizonicus*, on the other hand, prefers lower humidities (around 45 per cent), higher temperatures (20–25°C) and will generally move away from the light. Again, this accords well with the species' habits, since it lives in rather dry, warm places and is active at night. The reactions of the species change, however, and become harder to interpret when they are exposed to more extreme conditions. For instance, at high temperatures (35–40°C), *Armadillidium vulgare* tends to choose lower humidities, irrespective of whether these are in light or dark. One of several possible explanations for this behaviour is that, at these high temperatures, the species' physiological processes can be maintained only if body temperatures are lowered by permitting loss of water vapour from the body surface, which is more rapid at lower humidities. The normal reaction of *Venezilla arizonicus* changes if the species is exposed to very high humidities; it then tends to move to drier conditions even if these are in the light. Warburg concludes that, for these two species, light is not really an important physiological factor and acts mostly as a 'token stimulus', a clue to where optimum conditions of humidity and temperature may be found. For *V. arizonicus*, which lives in a dry or *xeric* habitat, darkness indicates the likely presence of the high temperatures and low humidities it prefers. For *Armadillidium vulgare*, in its cooler, more humid or *mesic* habitat, there is little risk of desiccation except in the most exposed situations and the species can afford to be relatively indifferent to light.

Warburg's study indicates the great complexity of the reactions of even relatively simple invertebrate animals such as Crustacea to the physical factors of their environment. If we analyse also the biotic factors of the animal's environment, food and enemies, the picture becomes even more complex. Other studies of *A. vulgare*, for instance, in California, indicate that the species shows quite strong preferences for different types of food (mostly various types of dead vegetation) and these also influence its distribution.

Knowing the physical requirements of a species can be of great economic significance, especially when an animal or plant is to be taken from one part of the world to another. The acacia trees of Australia, for example, are useful because many of them can grow in hot dry conditions and they can provide a valuable resource of fuel wood for human populations in the dry regions of the world. But there are many acacia species and each has its own peculiar climatic requirements. Trevor Booth and his colleagues at Canberra, Australia, have been documenting as many as possible of these requirements, based on the analysis of species distribution in Australia [27]. The climate of the region in which it is intended to introduce the acacia tree is then analysed and a computer match can be generated in which the ideal acacia for a particular region is selected. This has saved much wasted money and effort in avoiding the old trial-and-error style of forestry. Figure 3.21 shows those parts of Australia with a climate most similar to a proposed site for acacia introduction in Zimbabwe. The best acacia match for such requirements was found to be *Acacia holosericea*, the distribution map of which clearly corresponds very closely to this climatic requirement.

Species interaction

Physical factors evidently play an important part in determining the

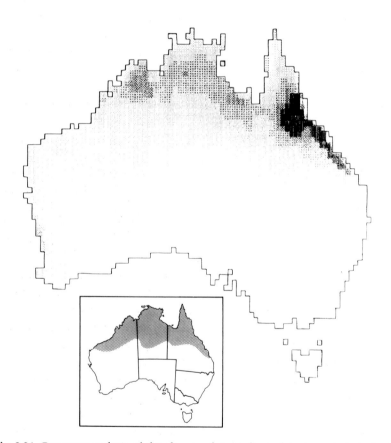

Fig. 3.21 Computer analysis of the climate of Australia to detect greatest similarity with that of Kadoma, Zimbabwe, which is to be the site of introduction of an acacia species for fuelwood production. Darkest areas have a climate most similar to that of the proposed site of introduction. Inset shows the distribution pattern in Australia of *Acacia holosericea*, a species which seems to match the climatic requirements for introduction. From Booth [27].

distribution limits of many plants and animals, but organisms also interact with one another, and this can also place constraints on geographical ranges. One species may depend strictly on another for food, as in the case of some butterflies which may be limited to a single food plant, or in the case of some host-specific parasites. Some species may be unable to colonize an area because of the existence of certain efficient predators or parasites in that area, or because some other species that is already established there can compete more efficiently for a particular resource that is in demand. These are *biotic factors*, and they are often responsible for limiting the geographical extent of a species within its potential physical range.

When a species is prevented from occupying an area by the presence of another species, this is termed *competitive exclusion*. It is not always easy to observe this ousting of one species by another in nature, but an example of its occurrence is in the barnacle species that occupy the rocky sea shores of western Europe and northeastern North America. Adult barnacles are firmly attached to rocks and feed by filtering plankton from the water when the tide is in. Two species that are common are *Chthamalus stellatus*, which is found within an upper zone of the shore just below high-tide mark, and *Balanus balanoides*, which occupies a much wider zone below that of *Chthamalus stellatus*, down to low-water mark. The distribution of the two species does not overlap by more than a few centimetres. This situation was analysed by the ecologist J.H. Connell [28], who found that when the larvae of *C. stellatus* ended their free-swimming existence in the sea and settled down for life, they did so over the whole upper part of the rocks above mean tide level. The larvae of *Balanus balanoides* settled over the whole zone between high and low water, including the area occupied by the adults of *Chthamalus stellatus*. Despite overlapping patterns of distribution of the larvae, different distributions of the adults of the two species result from two separate processes. One process acts on the zone at the top of the shore. The young *Balanus balanoides* are eliminated from this region because they cannot survive the long period of desiccation and the extremes of temperature to which they are exposed at low tide. *Chthamalus stellatus* are more resistant to desiccation and survive. Lower down the rocks the *Balanus balanoides* persist because they are not exposed for so long, and here the larvae of *Chthamalus stellatus* are eliminated by direct competition from the young *Balanus balanoides*. These grow much faster and simply smother the *Chthamalus* larvae or even prise them off the rocks. Connell also performed experiments on these species and found that, if adult *Balanus balanoides* were removed from a strip of rock and young ones prevented from settling, the *Chthamalus stellatus* were able to colonize the full length of the strip right down to low tide level. This showed that the competition with *Balanus balanoides* was the main factor limiting the distribution of *Chthamalus stellatus* to the upper part of the shore.

This example is a relatively simple one because only two species are involved. In most communities of animals and plants many species interact, and this makes it extremely difficult to sort out the full picture of the relationships between species. One approach to the problem is to remove one species from the community and to observe the reaction on the part of the others. This method has been tried in salt marsh communities in North Carolina by J.A. Silander and J. Antonovics [29], who removed selected plant species and recorded which of the other plants present in the community expanded into the spaces left behind (Fig. 3.22). They found a great range of responses. The removal of one grass species, *Muhlenbergia capillaris*, resulted in an equal expansion on the part of five other plants, suggesting that the grass was in competition with many other species. In the case of the sedge *Fimbristylis spadiceae*, however, removal led to the expansion of only one other plant, the chord grass species *Spartina patens*. The reciprocal experiment in which *Spartina* was removed similarly led to *Fimbristylis* taking full advantage of the new opportunity. In this case we seem to have only two species that are competing for this particular niche.

Selective removal of species in this way, however, is somewhat artificial and can result in disturbance to the physical environment that alters the very nature of the habitat, so it can only provide a preliminary guide to the relationships of species in the community.

There are many examples of the displacement of native species by an invader. The European starling, *Sturnus vulgaris*, for instance, was introduced into Central Park, New York, in 1891. Since then it has spread widely and is now found in all of the States (Fig. 3.23). It is mostly found in urban areas, and in the east has largely displaced the bluebird (*Sialia sialis*) and the yellow-shafted flicker (*Colaptes auratus*). These species nest in tree-holes or in man-made holes, and starlings can occupy and hold most of the limited supply of these nest sites. In the towns, then, the starling successfully competes with the native species for living space. But when flocks of starlings invade the countryside, they compete for food, insects and seeds with the meadow larks (*Sturnella*) and these birds also have declined in some areas. On the other hand, some North American species have been successful in new habitats in Europe and Australia. An example is the American grey squirrel (*Sciurus carolinensis*), which was introduced into the British Isles in the nineteenth century. Between 1920 and 1925 the native red squirrel (*S. vulgaris*) suffered a dramatic decline in numbers in Britain (largely due to disease following great abundance). The spread of the grey squirrel has been accompanied by the disappearance of the red squirrel from many areas, particularly those in which their numbers were reduced by disease and

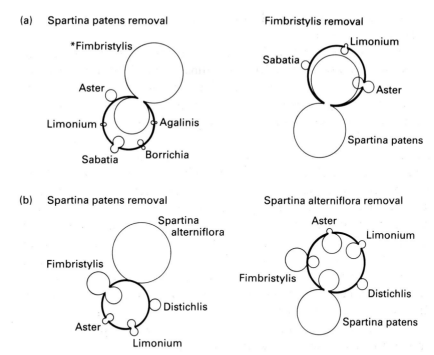

Fig. 3.22 A graphic illustration of the effect of removing a single plant species from a salt marsh community. The sizes of the circles represent the abundance of the plant species, and the heavy circle refers to the species that has been removed. Circles intruding into the heavy circle denote the response of different species to the perturbation. (a) High marsh site, where removal of *Spartina patens* results mainly in expansion of *Fimbristylis*, and removal of *Fimbristylis* results in expansion of *Spartina*. The two species seem to be in competition and the effects are roughly reciprocated. (b) Low marsh site where two species of *Spartina* predominate. Removal of *S. patens* results in no response by the other *Spartina* species, whereas removal of *S. alterniflora* does permit some expansion of *S. patens*. Competition here is not reciprocal. From Silander and Antonovics [29].

those in which the grey squirrel first spread and established itself. Where the grey squirrel has replaced the native red, it probably has done so by virtue of its superior adaptability to the niche of herbivore at canopy level in deciduous woodland.

There is good evidence that, in the past, whole faunas have invaded new areas and eliminated the native species by successful competition. For example, North and South America were separated by sea until the end of the Pliocene Period (about two million years ago). The Isthmus of Panama then came into being and enabled South American species to invade northwards and North American species southwards. In general, the northern species proved to be the more successful in competition, and most of the characteristic South American fauna of this time became extinct; but very few North American species were wiped out by the

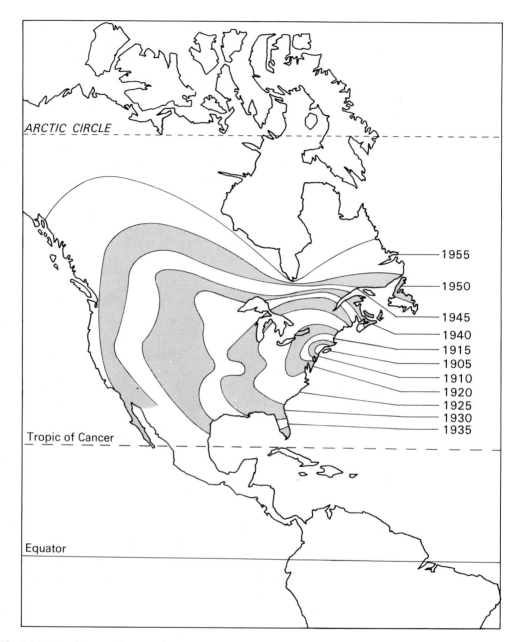

Fig. 3.23 Map showing the range extension of the European starling in North America following its introduction early in this century.

South American invaders — probably because the North American herbivores were more efficient and the predators more successful [30].

Despite these dramatic examples of invasion and competitive displacement, it is most likely that, in natural situations, species that compete for food or other resources have evolved means of reducing the pressures of competition and of dividing up the resources between them. This is mutually advantageous since it reduces the risk of either species being eliminated and made extinct by competition with the others. This is an advantage not only to the species directly involved but to the whole community of species in the habitat, since it results in *more* species, depending on as many different sources of food as possible. In such communities, competition occurs between so many different species, each with its own specialized adaptations, that no single species can become so numerous as to displace others. This may result in a greater degree of stability for the community, and stable communities are strongly resistant to the invasion of new species that might disrupt the highly evolved pattern of competition within them.

Reducing competition

Many different ways of reducing competition between species have evolved. Sometimes species with similar food or space requirements exploit the same resources at different seasons of the year, or even at different times of day. A common system amongst predatory mammals and birds is for one species (or a group of them) to have evolved specialized night-time activity whilst another species or group of species are day-time predators in the same habitats. An example amongst birds is the owls on the one hand, many species of which hunt at night, judging distance mostly by ear, and the hawks and falcons on the other, which are day-time hunters, with extremely keen eyesight, especially adapted for judging distances accurately. Thus, both groups of predators can coexist in the same stretch of country, and prey on the same limited range of small mammals. Cases of this sort are described as *temporal separation* of species, and this is an effective method of eking out limited food resources amongst several species. Among plants this process can be seen operating in deciduous forest habitats, where many woodland floor herbs flower and complete the bulk of their annual growth before the leaf canopy emerges on the trees. In this way the light-resources of the environment are used most efficiently.

A different type of temporal separation is shown within the complex grazing community of the East African savanna [31]. During the wet season all the five most numerous grazing ungulates (buffalo, zebra, wildebeest, topi and Thomson's gazelle) are able to feed together on the rich forage provided by the short grasses on the higher ground. At the beginning of the dry season, plant growth ceases there. The herbivores then descend to the lower, wetter ground in a highly organized sequence. First are the buffalo, which feed on the leaves of very large riverine grasses which are little used by the other species. The zebra, which are highly efficient at digesting the low-protein grass stems, move down next. By trampling the plants and eating the grass stems, they make the herb layer suitable for the next arrivals, topi and wildebeest. These two are found in slightly different areas. The jaws and teeth of the topi are adapted to the short, mat-forming grasses common in the north-western part of the Serengeti. Those of wildebeest are instead adapted to eating the leaves of the upright grasses commoner in the south-eastern Serengeti. These two species reduce the amount of grass, facilitating the grazing of the last species, Thomson's gazelle, which prefers the broader-leaved dicotyledonous plants to the narrow-leaved monocotyledonous grasses. The whole community, therefore, forms a complex ecosystem, which utilizes the pasture in a highly organized and efficient fashion.

Probably much more common, however, are cases where the resources of a habitat are divided up between species by the restriction of each of them to only part of it, to specialized microhabitats. This is called *spatial separation* of species; it means that each species must be adapted to live in the fixed set of physical conditions of its particular microhabitat. It also means that such a species is not adapted to live in other micro-habitats, and may find it difficult to invade them even if they were for some reason vacant and their food resources untapped.

An example of spatial separation has been described in the extensive marshlands of the Camargue in southern France, where different wading birds have different preferences for the various available feeding areas. The flamingo (*Phoenicopterus ruber*) has very long legs and is thus able to wade into deep water where it can sieve planktonic organisms with its highly specialized bill. In shallower water the avocet (*Recurvirostra avosetta*) and the shelduck (*Tadorna tadorna*) feed in a similar way, by sweeping, side-to-side actions of their necks. On the water's edge it is the Kentish plover (*Charadrius alexandrinus*) which feeds predominantly, being restricted to these regions by its shorter legs.

But the spatial patterns of distribution and feeding of predatory birds, such as these waders, sometimes reflect the patterns of their preferred food species. For example, the oystercatcher (*Haematopus ostralegus*) has a strong predilection for the bivalve mollusc *Cardium edule*, the cockle, and this is found mainly on sandy and muddy shores just below the mean high water mark of neap tides [32]. So this is the favourite feeding zone of the oystercatcher. Similarly, the mud-dwelling crustacean *Corophium volutator* is a favoured food species for the red-shank (*Tringa totanus*) and, since it thrives best in the upper regions of mudflats, usually above the mean high water mark of neap tides, this is often where large numbers of feeding redshanks can be found.

One can find separation of feeding behaviour even within a single species. In the oystercatcher, for example, even though the cockle is the main prey animal, individual birds within a flock may use different techniques for extracting the food from the protective cover of its shell

[33]. Some adopt a hammering technique, beating their bills against one of the valves until it breaks, while others use a more subtle approach, forcing their bills between the valves and prising them apart. Even among the hammerers there is further specialization, some attacking only the upper valve and others only the lower valve. These behavioural techniques are reflected in the bill structures of the individuals demonstrating such specializations. Careful observation of tagged birds has demonstrated that the hammerers have blunter bills than the stabbers (Fig. 3.24), but this is evidently an effect of bill wear and tear in the case of the hammerers, so there is no genetic basis for the separation of specialists. The advantage of this behaviour is that different oystercatchers are seeking prey with different weaknesses (either thin shells, or weak muscles holding the valves together), so at least some of the competition for food has been removed.

This type of feeding specialization within a species can under some circumstances lead to the development of distinctive races. In the arctic charr, a freshwater fish found in the lakes of mountainous regions of Europe, one may find two or more distinct forms in a single lake [34]. Figure 3.25 shows two forms of the charr discovered in Lake Rannoch in Scotland. The main difference is in the mouth structure. The upper fish has a relatively small mouth with small teeth. This animal is found to consume largely the zooplanktonic life found in the upper layers of the lake (the pelagic region). The lower fish has a larger, rounder mouth and bigger teeth. It eats a range of invertebrates and even fish, feeding mainly near the bottom (the benthic region). Perhaps this is the first stage in the evolution of two species from a single stock, the driving force being the advantage gained by avoiding competition for food.

This process has reached bizarre levels in Lake Victoria [35], where the number of forms of just 14 species of cichlid fish runs to 800. The sizes and shapes of these different forms are so varied that until very

Fig. 3.24 Bill size variation within one species, the oystercatcher (*Haematopus ostralegus*). The upper bird is a 'stabber', and extracts its shellfish prey by pushing the bill between the valves of the shell and prising it open. The lower bird is a 'hammerer' and breaks the shell by violent blows of the bill on one of the valves [33].

(a)

(b)

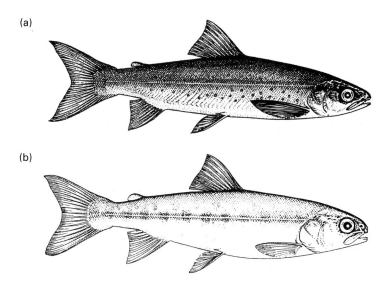

Fig. 3.25 Two different forms of the arctic charr from Loch Rannoch in Scotland. The upper fish has narrow jaws and small teeth, and feeds mainly on plankton. The lower fish has heavier jaws and larger teeth, and feeds on a wider range of prey, including invertebrates and other fish. From Walker, Greer and Gardner [34].

recently they were considered to be different species. Detailed study of their genetics has shown that they all, in fact, belong to a very restricted number of species. The avoidance of competition must be one of the factors leading to the evolution of new species and thus helping to increase the biodiversity of the planet.

It is this kind of development which has led to the high level of specialization found within some groups of species, such as the tanagers of Central America (Fig. 3.26). Three closely related species of tanager, the speckled tanager (*Tangara guttata*), the bay-headed tanager (*T. gyrola*) and the turquoise tanager (*T. mexicana*) may be found coexisting and feeding alongside one another without any competitive exclusion. The reason for this harmony is that each feeds in a slightly different location in the forest canopy. The speckled takes insects from the underside of leaves, the turquoise from fine twigs and the bay-headed from main branches. Each occupies its own niche and there is little overlap between them.

Even where overlap of niches does occur, animals may often coexist because they have their own distinctive location or way of life which they share with no other. This can be seen in Fig. 3.27, which displays diagrammatically the niches of various primate species in the tropical forest of Ghana [36]. Each species has its own preferred position in the forest canopy, some preferring undisturbed forest and others coping with exploited and cleared areas. There is considerable overlap in their

Fig. 3.26 Three species of tanager that coexist in the same forest on Trinidad. All feed on insects, but they exploit different microhabitats within the canopy and thus avoid direct competition. The speckled tanager (a) takes insects from the underside of leaves; the turquoise tanager (b) obtains its insects from fine twigs and leaf petioles; and the bay-headed tanager (c) preys upon insects on the main branches.

tolerances, but each has its own specialist location where it can hold its own.

Predators and prey

Predators may be another biological factor influencing the distribution of species, but their effects have been much less studied than those of competition. The simplest influences that predators might have is to eliminate species by eating them or, alternatively, to prevent the entry of new ones into a habitat. There is very little evidence that either of

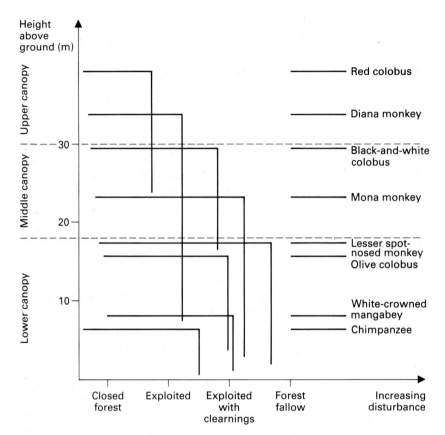

Fig. 3.27 Diagram to illustrate the different niche requirements of a range of primate species in the tropical forest of West Africa. Although the demands of the various species overlap, each has a particular height in the canopy or a type of site where it is most efficient and successful [36].

these processes is common in nature. One or two experimental studies have shown that predators sometimes eat all the representatives of a species in their environment, particularly when the species is already rare. But all such studies have been made in rather artificial situations in which a predator is introduced into a community of species that have reached some sort of balance with their environment in the absence of any predator; such communities are not at all like natural communities which already include predators. In general, it is not in the interests of predatory species to eliminate a prey species, because if they do this they destroy a potential source of food. Probably most natural communities have evolved so that there is a great number of potential prey species available to each predatory species. Thus no species is preyed upon too heavily, and the predators can always turn to alternative food species if the numbers of their usual prey should be reduced by climatic or other influences.

Prey switching of this type has been described on the island of Newfoundland, where the wolf and the lynx were major predators last century, but where the wolf is now extinct. The lynx was a rare animal until a new potential prey animal was introduced to the island in 1864 — the snowshoe hare. These multiplied rapidly, and so did the lynxes in response to the newly available food source. But the snowshoe hare population crashed to low levels in 1915 and the lynx, faced with starvation, switched its attentions to caribou calves, which had once been a major food source for the wolf. The snowshoe hare has now developed a 10-year cycle of high and low population levels and the lynx has continued to switch between hare and caribou depending upon whether the former is in a peak or a trough [37]. This pattern of prey switching allows the lynx to maintain a fairly stable population level and, as a consequence, it also permits the recovery of snowshoe hare populations.

This type of behaviour on the part of a predator may thus serve to prevent the extinction of its prey. Sometimes the relationship is even more complex; the predator may prevent the invasion of more efficient or voracious predators which could reduce the prey population yet further. A good example of this is the Australian bell miner (*Manorina melanophrys*) [38]. This is a communal and highly territorial bird which occupies the canopy of eucalypt forests and which feeds largely on the nymphs, secretions and scaly covers of psyllid plant bugs. Yet the psyllid bugs survive well under this predation and, what is even more remarkable, they seem to require the attentions of the bell miners, for when these birds are removed, the populations of bugs crash and the eucalypt trees become more healthy. It seems that the aggressive behaviour of the miners towards other birds prevents their entering bell miners' territories and eating all the psyllids. While the bell miners stay the psyllids are safe, but the trees suffer!

As mentioned earlier, competition may prevent two species from living together in a habitat, and may modify the distribution of species, because the resources of the habitat are inadequate to support both of them. Probably the most important effect of predators and of parasites and disease (which are 'internal' predators) on the distribution of species is that, by feeding on the individuals of more than one species, they reduce the pressures of competition between them. Thus, by reducing pressures on the resources of the habitat, predators may allow more species to survive than would survive if the predators were not there. It has been shown by laboratory experiments that, if two species of seed-eating beetle (weevils) were kept together in jars of wheat, one species always eliminated the other within five generations. One of the two species always multiplied faster than the other, and this species won in the competition for food and places to lay eggs. But if a predator was introduced, such as a parasitic wasp, whose larvae feed inside the bodies of the beetle larvae of either species and eventually kill them, both species persisted. The numbers of both species were kept so low by the predator that competition for food, which would have caused one or other to be eliminated, never occurred.

Other studies of natural communities have largely confirmed the hypothesis that predators may actually *increase* the number of different species that can live in a habitat. The American ecologist Robert T. Paine made an especially fine study on the animal community of a rocky shore on the Pacific coast of North America [39]. The community included 15 species, comprising acorn barnacles, limpets, chitons, mussels, dog whelks and one major predator, the starfish *Pisaster ochraceus*, which fed on all the other species. Paine carried out an experiment on a small area of the shore in which he removed all the starfish and prevented any others from entering. Within a few months 60–80 per cent of the available space in the experimental area was occupied by newly settled barnacles, which began to grow over other species and to eliminate them. After a year or so, however, the barnacles themselves began to be crowded out by large numbers of small, but rapidly growing mussels, and when the study ended these completely dominated the community, which now consisted of only eight species. The removal of predators thus resulted in the halving of the number of species and there was evidence, too, that the number of plant species of the community (rock-encrusting algae) was also reduced, because of competition with the barnacles and mussels for the available space.

A general conclusion, then, is that the presence of predators in a well-balanced community is likely to increase rather than reduce the numbers of species present, so that, overall, predators broaden the distribution of species. Only a few experiments similar to Paine's have been performed and so one must be cautious about applying this conclusion to all communities. There is some independent evidence, however, that herbivores, which act on plants as predators do on their prey, may similarly increase the number of plant species that can live in a habitat. In the last century, Charles Darwin noticed that in southern England, meadowland grazed by sheep often contained as many as 20 species of plants, while neglected, ungrazed land contained only about 11 species. He suggested that fast-growing, tall grasses were controlled by sheep grazing in the meadow, but that in ungrazed land these species grew tall so that they shaded the small slow-growing plants from the sun and eliminated them. A similar process occurred in chalk grassland areas in Britain, when the disease myxomatosis caused the death of large numbers of rabbits; the resulting reduction in grazing allowed considerable invasion by coarse grasses and scrub. As a result many of these areas are much less rich in species than they were under heavy 'predation'.

On the Washington coast, Paine performed another series of experiments in which he removed the sea-urchin *Strongylocentrotus purpuratus*, which grazes on algae [40]. Initially, there was an increase in the number of species of algae present; the six or so new species were probably ones that were normally grazed too heavily by the sea-urchin to survive in the habitat. But over two or three years the picture changed as the community of algae gradually became dominated by two species, *Hedophyllum sessile* on exposed parts of the shore, and *Laminaria groenlandica* in the more sheltered regions below low-water mark. These two species were tall and probably 'shaded out' the smaller species, as did the tall grasses studied by Darwin. The total number of species present was in the end greatly reduced after the removal of the herbivores.

The activities of carnivorous predators in a community also have an effect on the plants since, by limiting to some extent the number of their herbivorous prey, they prevent overgrazing, and thus reduce the risk of rare species of plants being eliminated.

The interactions may be quite complex, however, as in the case of the Hawaiian damselfish, which is a predator of coral reel habitats. In an experimental study of the influence of this fish [41], plates were constructed which were suitable for algal colonization, and these were placed in three types of location: (i) within cages which excluded all herbivorous fishes, (ii) uncaged, but within the territories of the carnivorous damselfish and (iii) uncaged and placed outside damselfish territories. The diversity of the colonizing algae was highest on the uncaged plates inside damselfish territories and least in the uncaged samples outside the territories. In other words, where there was no grazing at all the algal diversity was higher than when there was intense grazing, but the highest diversity was found in sites where grazing was controlled to an extent by the predation of the damselfish upon the grazers.

The complicated sets of interactions between predator, grazer and plant can lead to the development of a finely balanced and diverse ecosystem, as shown by this coral reef example.

References

1 Barton, N.H. (1988) Speciation. In *Analytical Biogeography* (eds A.A. Myers & P.S. Giller). Chapman & Hall, London, pp. 185–218.

2 Rosen, B.R. (1988) Biogeographic patterns: a perceptual overview. In *Analytical Biogeography* (eds A.A. Myers & P.S. Giller). Chapman & Hall, London, pp. 23–55.

3 Whitmore, T.C. (1978) The palm family. In *Flowering Plants of the World* (ed. V.H. Heywood). Oxford University Press, Oxford, pp. 301–304.

4 Sharrock, J.T.R. & Sharrock, E.M. (1976) *Rare Birds in Britain and Ireland*. T. & A.D. Poyser, Berkhamsted.

5 Sagar, G.R. & Harris, J.L. (1964) Biological flora of the British Isles, *Plantago major* L., *P. media* L, and *P. lanceolata* L. *Journal of Ecology* **52**, 189–221.

6 Corbet, P.S., Longfield, C. & Moore, N.W. (1960) *Dragonflies*. Collins, London.

7 Corbet, P.S. (1957) The life history of the emperor dragonfly, *Anax imperator* Leach (Odonata: Aeshnidae). *Journal of Animal Ecology* **26**,1–69.

8 Dandy, J.E. (1981) Magnolias. In *The Oxford Encyclopedia of Trees of the World* (ed. B. Hora). Oxford University Press, Oxford, pp. 112–114.

9 Cassagnau, P. (1959) Contribution a la connaissance du genre *Tetracanthella* Schott. *Memoires de la Musee d'Histoire Naturelle de Paris* **16**(7),201–260.

10 Corbet, G.B. & Southern, H.N. (eds) (1977) *The Handbook of British Mammals*, 2nd edn. Blackwell Scientific Publications, Oxford.

11 Coope, G.R. (1973) Tibetan species of dung beetle from Late Pleistocene deposits in England. *Nature* **245**,335–336.

12 Hamilton, A.C. (1982) *Environmental History of East Africa*. Academic Press, London.

13 Favarger, C. (1972) Endemism in the montane floras of Europe. In *Taxonomy, Phytogeography and Evolution* (ed. D.H. Valentine). Academic Press, London, pp. 191–204.

14 Meyer, E.F. (1973) Late Quaternary paleoecology of the Cuatro Cienegas Basin, Coahuila, Mexico. *Ecology* **54**,982–995.

15 Lewis, H. (1972) The origin of endemics in the California flora. In *Taxonomy, Phytogeography and Evolution* (ed. D.H. Valentine). Academic Press, London, pp. 179–189.

16 Kendeigh, S.C. (1974) *Ecology with Special Reference to Animals and Man*. Prentice Hall, Englewood Cliffs New Jersey.

17 Marshall, J.K. (1978) Factors limiting the survival of *Corynephorus canescens* (L.) Beauv. in Great Britain at the northern edge of its distribution. *Oikos* **19**,206–216.

18 Root, T. (1988) Energy constraints on avian distributions. *Ecology* **69**,330–339.

19 Thompson, P.A. (1978) Germination of species of Caryophyllaceae in relation to their geographical distribution in Europe. *Annals of Botany* **34**,427–449.

20 Moore, P.D. (1981) The varied ways plants tap the sun. *New Scientist* **81**,394–397.

21 Edwards, G. & Walker, D.A. (1983) C_3, C_4: *Mechanisms and Cellular and Environmental Regulation of Photosynthesis*. Blackwell Scientific Publications, Oxford.

22 Ehleringer, J.R. (1978) Implications of quantum yield differences on the distribution of C_3 and C_4 grasses. *Oecologia* **31**,255–267.

23 Teeri, J.A. & Stowe, L.G. (1976) Climatic patterns and the distribution of C_4 grasses in North America. *Oecologia* **23**,1–12.

24 Williams, G.J. & Markley, J.L. (1973) The photosynthetic pathway type of North American shortgrass prairie species and some ecological implications. *Photosynthetica* **7**,262–270.

25 Spooner, G.M. (1974) The distribution of *Gammarus* species in estuaries. *Journal of the Marine Biological Association* **27**,1–52.

26 Warburg, M.R. (1968) Behavioural adaptations of terrestrial isopods. *American Zoologist* **8**,545–599.

27 Booth, T. (1988) Which wattle where? Selecting Australian acacias for fuelwood plantations. *Plants Today* **1**,86–90.

28 Connell, J. (1961) The influence of interspecific competition and other factors on the distribution of the barnacle *Chthamalus stellatus*. *Ecology* **42**,710–723.

29 Silander, J.A. & Antonovics, J. (1982) Analysis of interspecific interactions in a coastal plant community — a perturbation approach. *Nature* **298**,557–560.

30 Webb, S.D. (1978) A history of savannah vertebrates in the New World. Part II: South America and the Great Interchange. *Annual Reviews of Ecology and Systematics* **9**,393–426.

31 Bell, R.H.V. (1970) The use of the herb layer by grazing ungulates in the Serengeti. In *Animal Populations in Relation to their Food Resources* (ed. A. Watson). Blackwell Scientific Publications, Oxford, pp. 111–127.

32 Hale, W.G. (1980) *Waders*. Collins, London.

33 Sutherland, W.J. (1987) Why do animals specialize? *Nature* **325**,483–484.

34 Walker, A.F., Greer, R.B. & Gardner, A.S. (1988) Two ecologically distinct forms of Arctic Charr *Salvelinus alpinus* (L.) in Loch Rannoch, Scotland. *Biological Conservation* **43**,43–61.

35 Meyer, A., Kocher, T.D., Basasibwaki, P. & Wilson, A.C. (1990) Monophyletic origin of Lake Victoria cichlid fishes suggested by mitochondrial DNA sequences. *Nature* **347**,550–553.

36 Martin, C. (1991) *The Rainforests of West Africa*. Birkhauser Verlag, Basel.

37 Bergerud, A.T. (1983) Prey switching in a simple ecosystem. *Scientific American* **249**(6),116–124.

38 Lyon, R.H., Runnalls, R.G., Forward, G.Y. & Tyers, J. (1983) Territorial bell miners and other birds affecting populations of insect prey. *Science* **221**,1411–1413.

39 Paine, R.T. (1966) Food web complexity and species diversity. *American Naturalist* **100**,65–75.

40 Paine, R.T. & Vadas, R.L. (1969) The effect of grazing in the sea urchin *Strongylocentrotus* on benthic algal populations. *Limnologia Oceanographica* **14**,710–719.

41 Hixon, M.A. & Brostoff, W.N. (1983) Damselfish as keystone species in reverse: intermediate disturbance and diversity of reef algae. *Science* **220**,511–513.

4 Communities, ecosystems and biomes

The fact that species interact with one another means that no species can be fully understood in terms of its ecology and geographical distribution in isolation from other species. The range of predators and herbivores is influenced by that of their prey or food organisms. Parasites are severely limited by their hosts and some plants may even be limited by their disperser animals or by pollinators. Added to these interactions, many organisms overlap in their physical or resource requirements and may therefore be found together in the same locations. The outcome is that some plant and animal species tend to be associated with one another, and this has led to the development of the concept of a community of different species, each to some extent dependent on its neighbours either directly or indirectly. Thus some plant species of a forest floor may be partially or completely parasitic on a host tree, in which case their relationship is close and the presence of the parasite is dependent on the presence of the host. Alternatively, independent photosynthetic plants may not be so directly related to the trees with which they are associated, but may require the conditions of shade, moderate temperature and humidity that can be provided only by a certain composition of canopy. Their soil requirements may also result in an apparent association with certain forest trees, thus lending weight to the idea of a plant community.

Some animal species, such as those herbivores with very specific food requirements, may form close, dependent unions with certain plants, but others may have requirements for those spatial architectural conditions that are best supplied by particular assemblages of plants. Again, the outcome of such associations is the existence of communities of plants and animals in nature which, within a geographic area, may be repeated in similar topographic and environmentally comparable sites, and which may be very predictable in their species composition [1].

The idea of discrete communities that can be described, and even named and classified [2], is very attractive to the human mind and is certainly valuable in the process of mapping areas and assessing their value for conservation and in determining plans for management. But it must be recognized that nature probably does not consist of a series of such communities that operate as distinct entities. Very rarely do the distribution ranges of any two species overlap precisely, and the degree of association between ground flora and canopy, or between invertebrate animal assemblages and vegetation, is often weaker than one might assume from casual observation. Studies of the past history of plant and animal species over the last 10 000 years or so have also demonstrated that they come into contact at certain times in their history, but have also been periodically separated as the climate changed. Often the associations we now observe are of relatively recent origin and should be regarded as transitory. The concept of the community must, therefore, be looked upon as useful but somewhat artificial.

The ecosystem

An alternative way of looking at the organization of assemblages of plants and animals is to regard them, together with the non-living components of the environment, as integral parts of an interactive whole, called the ecosystem. Whereas the idea of community concentrates on the different species found in association with one another, the concept of the ecosystem is largely concerned with the processes that link different organisms to one another, rather than with the regularity with which certain species are associated.

There are two fundamental ideas that underlie the ecosystem concept; these are *energy flow* and *nutrient cycling*. Energy, initially fixed from solar radiation into a chemical form by green plants, moves into herbivores as a result of their feeding upon plants, and then moves on into carnivores as the herbivores are themselves consumed. Since herbivores rarely consume all the available plant material, and since carnivores do not eat every morsel of their prey organisms, some living tissues are allowed to die before being eaten. These dead tissues constitute an energy resource which can be exploited by scavengers, detritus feeders and decomposers. Thus, we can classify all the organisms in any community in terms of their feeding behaviour. In practice, a series of complex *feeding webs* are usually formed, relating each species to many others, whether as feeder or food.

Energy is the ability to do work. It can be measured in a mechanical form, where the *erg* is the energy needed to raise one milligram through one centimetre. The *joule* is 10^7 ergs and the kilojoule (kJ) is 10^3 joules.

Figure 4.1 shows the course and the rate of energy movement through an ecosystem in which sheep are acting as the major grazer in a mountain grassland habitat [3]. Of the total solar energy arriving at the site $(3 \cdot 08 \times 10^6 \, \text{kJ}^{-2} \, \text{yr}^{-1})$, only about 40 per cent is available to the plants. The remainder cannot be absorbed by chlorophyll, because it is of the wrong wavelength. Of the energy fixed in photosynthesis, a proportion is lost in the respiration of the plant, so that ultimately $21\,158 \, \text{kJ} \, \text{m}^{-2}$ is built up in the above-ground plant tissues and $6053 \, \text{kJ} \, \text{m}^{-2}$ in the roots each year. The sum of these two, $27\,211 \, \text{kJ} \, \text{m}^{-2} \, \text{yr}^{-1}$, represents the *net primary production* of the system (net in the sense that plant respiratory losses have been subtracted). This rate of production accounts for only

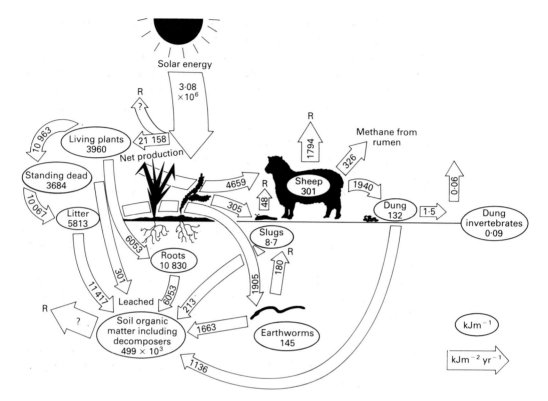

Fig. 4.1 Energy in a sheep-grazed, grassland ecosystem in Wales (data from Perkins [3]). Units are kJ m^{-2}, or kJ m^{-2} yr^{-1} if within an arrow.

about 2 per cent of the energy actually available for photosynthesis (0·8 per cent of the total incident energy), so the overall efficiency of the process of primary production is very low. Estimates for the entire global efficiency of primary production range from 0·07 per cent for the oceans to 0·3 per cent as a mean value for all land communities.

Despite this production rate, the amount of energy residing within the vegetation at any given time is only 3960 kJ m^{-2}, which means that there must be a rapid turnover within the system. Part of this turnover is accounted for by the sheep grazing, which consumes 4659 kJ m^{-2} yr^{-1} (about 22 per cent of the above-ground primary production) and the remainder dies and goes into litter production. Almost 20 per cent of the energy finding its way into this litter is consumed by earthworms, the remainder being incorporated into the soil by a variety of other routes.

When vegetable matter passes through the gut of an animal, not all of the energy is used efficiently, so that faecal material still forms an energy resource for other organisms. The energy supply in sheep dung represents 1940 kJ m^{-2} yr^{-1}, some of which is exploited by dung beetles and other dung-feeding invertebrates, but the bulk of it finds its way into the decomposer food chains in the soil. Overall, it can be seen that there is considerably more energy both stored within and moving through the soil/decomposer part of the ecosystem than there is in the grazing food chains above the ground surface. This is true of many ecosystems.

It is easy to underestimate the abundance of detritivores and decomposers in an ecosystem, simply because they are small and often live out of sight in litter or in the soil. It has been estimated that a square metre of woodland soil can contain 1000 species of invertebrate animals, and in such an area the population of nematode worms may exceed 10 million. Earthworms in this square metre probably number 300−500, and bacteria flourish within their guts, so that the excreta of earthworms may contain densities of bacteria 200−400 times as great as those in the ingested litter. All of this mass of decomposer organisms is converting organic carbon compounds into carbon dioxide gas and, as it respires, liberating energy in the form of heat.

For this particular ecosystem we know nothing of the energy flow through the carnivores in the system. For example, slugs, earthworms and dung invertebrates, as well as other invertebrate grazers, will be eaten by ground-feeding birds, such as the meadow pipit, or by mammals such as shrews. These in turn may be preyed upon by 'top carnivores' such as fox or kestrel. These various feeding levels are termed *trophic levels*, for example:

Trophic level   1   2   3   4

Plant → Herbivore → Carnivore   → Top carnivore
Grass   Crane fly   Meadow pipit   Kestrel
             larvae

One may find extra trophic levels in a food chain, for example:

Grass → Crane fly → Carnivorous   → Shrew → Snake → Fox
             larvae   ground beetle

but food chains of this length are unusual.

Nutrients in the ecosystem

While energy flows through the ecosystem between trophic levels, the chemical elements, of which matter is composed, cycle within the system. Some elements such as carbon and oxygen, both of which are common in gaseous compounds, move freely between ecosystems, whereas others such as phosphorus, calcium and potassium do not have common gaseous forms and are therefore less mobile. Even these, however, enter and leave ecosystems by various means. For example, precipitation brings various elements into ecosystems in both dissolved and

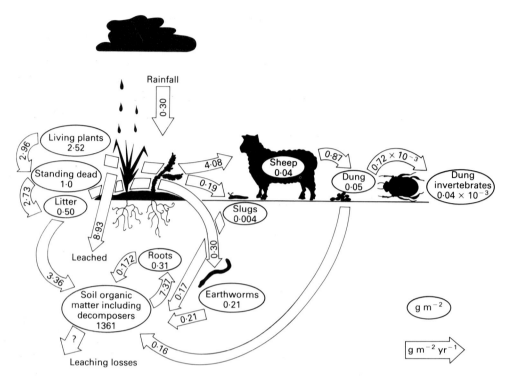

Fig. 4.2 Movement and storage of potassium within a sheep-grazed upland grassland ecosystem in Wales (data from Perkins [3]). Units are $g\,m^{-2}$, or $g\,m^{-2}\,yr^{-1}$ if figures are within arrows.

suspended forms. In Fig. 4.2, which portrays the same sheep-grazed mountain grassland ecosystem as in Fig. 4.1, the amount of potassium entering the ecosystem in rainfall is as high as $0.3\,g\,m^{-2}\,yr^{-1}$. This high input is due to the fact that the mountains of North Wales, from which these figures were obtained, are close to the sea and the rainfall is therefore rich in a variety of elements.

Other gains to the ecosystem may result from animal migrations, the carriage of plant matter by wind or water, or by human fertilization to improve productivity. The weathering of rock materials within the soil is another source of elements for the ecosystem. The ecosystem also suffers losses, the most important of which is usually the leaching of chemicals from the plants, litter and soil as water moves through the system and out in stream flow or by soaking away to underground aquifers. It is very difficult to obtain reliable values for the rate of weathering in the soil, but it is likely to be of the same order as the leaching losses. The overall budget for an ecosystem can be given by the following equation:

Soil weathering + Nutrient input from rainfall, etc.

= Leaching loss + Other losses + Growth in biomass

Within the ecosystem a given atom may be recycled many times before being lost. It will be moving constantly from *abiotic* (non-living) to *biotic* (living) materials and back again. Its incorporation into biotic material, as for example when it is absorbed by a plant root, involves the expenditure of energy on the part of the plant to pump it across the cell membrane. So some of the respiratory energy losses from the ecosystem are used in driving the nutrient cycles.

The distribution and flow rate of an element within an ecosystem may be of considerable interest. In the grassland system there is five times the concentration of potassium per unit area in earthworms than in sheep. Indeed, there is more potassium in sheep dung than in sheep. It is also interesting to note that the transfer of potassium to the sheep in grazing exceeds the rate of movement of potassium into the litter following death of plant tissue. This contrasts with the energy situation (Fig. 4.1).

A knowledge of the quantities and flow rates of such ecosystems helps us to manage them efficiently. If we wish to crop the ecosystem at any given trophic level (e.g. sheep at the second trophic level), then we must ensure that the rate of removal of nutrients can be compensated for by natural inputs from rainfall and weathering. If this is not so, then we must either reduce the level of exploitation or add those elements in deficit as fertilizers. An element in short supply may limit the rate of ecosystem productivity.

Global communities

The application of the community concept on a global scale is complicated by the restriction of species to certain geographic areas. This means that any system of classifying the world's vegetation into units must avoid the simple use of particular groups of species. Although the rain forests of Brazil are comparable in many respects to those of West Africa or South-East Asia, the actual assemblage of species is quite different. Their comparability is mainly due to the fact that they are structurally similar, dominated by tall trees arranged in a series of layers. Such a structure can be achieved only under specific climatic conditions, with high temperatures and with a rich supply of rain throughout the year. Only the equatorial parts of the globe can satisfy these conditions. In order to understand global communities of plants and animals, therefore, it is necessary to consider the pattern of climate over the earth's surface.

The climate of an area is the whole range of weather conditions, temperature, rainfall, evaporation, water, sunlight and wind that it experiences through all the seasons of the year. Many factors are involved in the determination of the climate of an area, particularly latitude, altitude, and position in relation to seas and land masses. The climate in turn largely determines the species of plants and animals that can live in an area.

Climate varies with latitude for two reasons. The first reason is that the spherical form of the earth results in an uneven distribution of solar energy with respect to latitude. As the angle of incidence of the sun's rays approaches 90°, the area over which the energy is spread is reduced, so that there is an increased heating effect. In the high latitudes, energy is spread over a wide area; thus polar climates are cold (Fig. 4.3(a)). The precise latitude that receives sunlight at 90° at noon varies during the year; it is at the Equator during March and September, at the Tropic of Cancer (23·45°N) during June, and at the Tropic of Capricorn (23·45°S) during December. The effect of this seasonal fluctuation is more profound in some regions than in others.

The second reason is that variations also result from the pattern of movement of air masses (Fig. 4.3(b)). Air is heated over the Equator, and therefore rises (causing a low-pressure area) and moves towards the pole. As it moves towards the pole it gradually cools and increases in density until it descends, where it forms a subtropical region of high pressure, known as the Horse Latitudes (Fig. 4.3(c)). Air from this high pressure area either moves towards the Equator, or else moves polewards. This latter air eventually meets cold air currents moving south from the polar region, over which air is cooled and descends (causing a high-pressure area). Where these two air masses meet, a region of unstable low pressure results, in which the weather is changeable [4].

This idealized picture is complicated by the *Coriolis effect* (named in honour of the French mathematician Gaspard Coriolis, who analysed it), which results from the west–east rotation of the earth. This force tends to deflect a moving object to the right of its course in the Northern Hemisphere and to the left in the Southern Hemisphere. As a result, the winds moving towards the Equator come to blow from a more easterly direction (Fig. 4.3(c)). These 'Trade Winds', coming from both the Northern and the Southern Hemispheres, therefore meet at the Equator, and this region is known as the 'inter-tropical convergence'. Where these easterly winds have passed over oceans they have become moist, and this moisture is deposited as rain over the easterly portions of the equatorial latitudes of the continents. Similarly, the winds which move polewards from the high-pressure Horse Latitudes come to blow from a more westerly direction, and provide rain along the westerly regions of the higher latitudes of the continents. The Horse Latitudes themselves are regions in which dry air is descending, and arid belts form along these latitudes of the continents.

The distribution of oceans and land masses modifies this simple picture yet further. Because heat is gained or released more slowly by water than by land masses, heat exchange is slower in maritime regions, while at the same time humidities are higher. In summer, therefore, continental areas tend to develop low-pressure systems as a result of the heating of land masses and the conduction of this heat to the overlying air masses. Conversely, in winter the reverse situation occurs, continental areas becoming cold faster than the oceans, and high-pressure systems develop over them (Fig. 4.3(d) and (e)). One effect of this process is that the continental low-pressure systems draw in moist air from neighbouring seas (for example, from the Indian Ocean to East Africa and India) causing summer monsoon rains. The winter of these areas, on the other hand, is usually dry.

In addition to the heating and cooling effects of land masses, climate is also affected by altitude. On average, the air temperature falls by 0·6°C for every 100 m rise in height, but this varies considerably according to prevailing conditions, especially the aspect and steepness of slope and the wind exposure. Because of this tendency for temperature to fall with increasing altitude, the organisms inhabiting high tropical and subtropical mountains — such as the Himalayas in northern India (Fig. 4.4) — may be more like the flora and fauna of colder regions than that of the surrounding lowlands. However, although temperature in general falls as one ascends such mountains, other environmental conditions do not precisely mirror those found at higher latitudes. For example, the seasonal variations in day-length typical of high-latitude tundra areas are not found in the 'alpine' regions of tropical mountains. Also the high degree of insolation resulting from the high angle of the sun produces considerable diurnal fluctuations in temperature that are not found in tundra regions. It is not surprising, therefore, that the altitudinal zonation of plants and animals should not precisely reflect the global, latitudinal zonation. Also, arctic and alpine races of a single species often differ in their physiological make-up as a consequence of these climatic differences.

Climate diagrams

As we have seen, plants and animals are affected by a whole range of physical factors in their environment, many of these being directly related to climate. Biogeographers have long sought, therefore, a means of portraying climates in simple, condensed form that would give at a glance an indication of the main features that might be of critical importance to the organisms of the area. Mean values of temperature and rainfall may be of some use, but one also needs to know something of seasonal variation and of extreme values if the full implications of a

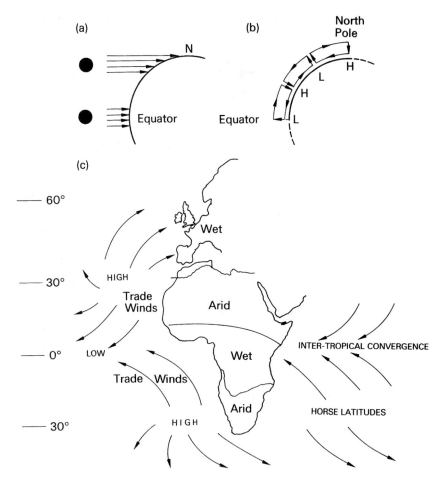

Fig. 4.3 Patterns of climate. (a) Due to the spherical shape of the earth polar regions receive less solar energy per unit area than the equatorial regions. (b) The major patterns of circulating air masses (cells) in the Northern Hemisphere: H, high pressure; L, low pressure. (c) Pressure areas, wind directions and moisture belts around Europe and Africa.

particular climatic regime are to be appreciated. It is with this aim in view that Heinrich Walter, of the University of Hohenheim in Germany, devised a form of climate diagram which is now widely used by biogeographers [5]. An explanation of the form of these diagrams is given in Fig. 4.5, and they are used later in this chapter in displaying the climates of the major biological zones of the earth.

These diagrams are particularly effective at emphasizing those aspects of climate that are most important in determining the type of vegetation that can survive in an area and the structure which that vegetation will assume. The characteristic assemblage of plants and animals that develops upon this climatic template is termed the *biome* (e.g. tropical rain forest, coniferous forest, tundra, etc.). The distinctions between

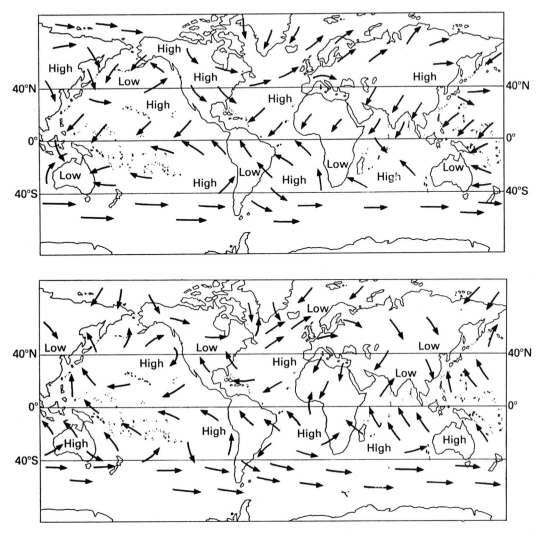

Fig. 4.3 (d and e) General pattern of air movement across the earth's surface and distribution of areas of high and low pressure (d) in January and (e) in July. Note the reversal of winds in the Indian Ocean in July, taking monsoon rains into northern India and East Africa.

biomes are not necessarily related to the taxonomic classification of the organisms they contain, but rather to the *life-form* (the form, structure, habits and type of life history of the organism in response to its environment) of their plants and animals. This concept of the life-form was first put forward by the Danish botanist Christen Raunkaier in 1903. He observed that the most common or dominant types of plants in a climatic region had a form well suited to survive in the prevailing conditions.

Thus in arctic conditions the most common plants are dwarf shrubs and cushion-forming species that have their buds close to ground level.

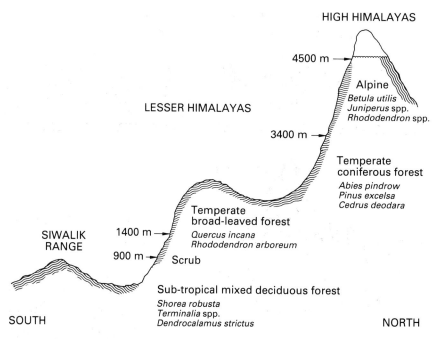

Fig. 4.4 Diagrammatic section of the western Himalayas in northern India, showing the approximate altitudinal limits of the major vegetation types. With increasing altitude one passes through vegetation belts similar to those found on passing into higher latitudes. The scrub zone (900–1400 m) is strongly modified by human deforestation.

In this way they survive the winter conditions when wind-borne ice particles would have an abrasive effect on any elevated shoots. In warmer climates, buds may be carried well above the ground and the tree is an efficient life-form, but periodic cold or drought may necessitate the loss of foliage and the development of a dormant phase. This has resulted in the evolution of the deciduous habit. Prolonged drought is again associated with shrubs that have smaller above-ground structures. Some plants of areas with seasonal drought survive the unfavourable period as underground perennating organs (bulbs or corms) or as dormant seeds. Animals also show distinct life-forms adapted to different climates with cold-resistant, seasonal or hibernating forms in cold regions and forms with drought-resistant skins or cuticles in deserts. Nevertheless, animal life-forms are usually far less easy to recognize than are those of plants and, consequently, most biomes are distinguished by the plants they contain and are named after their dominant plant life-form.

There is no real agreement among biogeographers about the number of biomes in the world. This is because it is often difficult to tell whether a particular type of vegetation is really a distinct form or is merely an early stage of development of another, and also because many types of vegetation have been much modified by the activities of man.

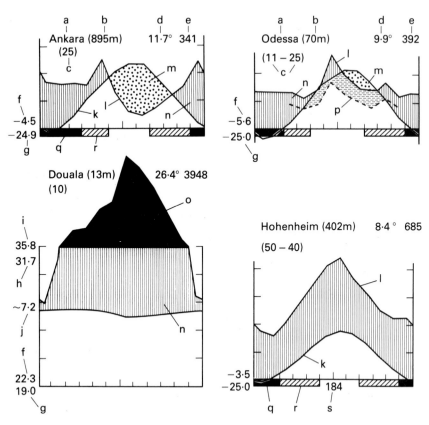

Fig. 4.5 Climate diagrams convey many aspects of seasonal variation in climate at a site in a manner that is easily assimilated visually, and which is of great relevance to the determination of vegetation in the area. Key to the climatic diagrams. Abscissa: months (N. Hemisphere January–December, S. Hemisphere July–June); Ordinate: one division = 10°C or 20 mm rain. a = station, b = height above sea-level, c = duration of observations in years (of two figures the first indicates temperature, the second precipitation), d = mean annual temperature in °C, e = mean annual precipitation in mm, f = mean daily minimum of the coldest month, g = lowest temperature recorded, h = mean daily maximum of the warmest month, i = highest temperature recorded, j = mean daily temperature variations, k = curve of mean monthly temperature, l = curve of mean monthly precipitation, m = relative period of drought (dotted), n = relative humid season (vertical shading), o = mean monthly rain > 100 mm (black scale reduced to 1/10), p = reduced supplementary precipitation curve (10°C = 30 mm) and above it (dashes) dry period, q = months with mean daily minimum below 0°C (diagonal shading), r = late or early frosts occur, s = mean duration of frost-free period in days. Some values are missing, where no data are available for the stations concerned (h−j are only given for diurnal types of climate). After Walter [5].

We shall describe eight terrestrial biomes, a freshwater one, and three marine ones. Figure 4.6 shows the ways in which these biomes are related to climatic factors and Fig. 4.7 shows their geographical distributions.

The distribution of biomes shown in Fig. 4.7 is somewhat idealized

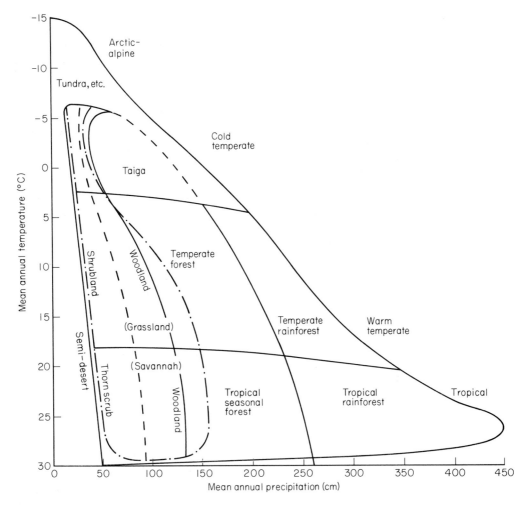

Fig. 4.6 The distribution of the major terrestrial biomes with respect to mean annual precipitation and mean annual temperature. Within regions delimited by the dashed line a number of factors, including oceanicity, seasonality of drought and human land-use may affect the biome type which develops. From Whittaker [6]. Reprinted with the permission of Macmillan Publishing Company from *Communities and Ecosystems,* Second Edition by Robert H. Whittaker. Copyright © 1975 by Robert H. Whittaker.

because human impact has often severely modified the pattern of biomes over the earth and has also destroyed many of the characteristic vegetation types and replaced them with agricultural land or cities. It has become quite unrealistic to consider biomes in the absence of such human impact, so the brief accounts given here of specific biomes will emphasize the problems that the natural systems face under increasing human pressures.

A further consideration, also related to the influence of our species on global ecology, is the fact that the climate is changing, so the boundaries we conveniently draw around the biomes on maps such as Fig. 4.7 can be expected to change in response to rising global temperature. Theoretical studies based on predictive models of rising carbon dioxide and the development of a 'greenhouse effect' have been extensively used to estimate the future trends in global temperature, but the models are by their very nature complex and many variables are still unknown. Although much has been written about the gases entering our atmosphere that will enhance its heat retention, little is yet known about the effects of other gaseous products of human industrial activity, such as sulphur dioxide, which may lead to greater radiative losses from the atmosphere [7]. While such basic questions as these remain in doubt, it is impossible to make firm predictions about the future of our climate. It is now firmly established that global temperatures have been rising for the last 150 years, having increased by about 1°C in that time, and there is some evidence that the rate of rise is itself increasing, with about 0·5°C gained in the last 50 years (see Fig. 11.14). But climate varies for a number of different reasons and it is, as yet, impossible to be sure that the current trend is not just part of the background fluctuations. Until the temperature trend can be firmly established it will not be possible to estimate accurately the outcome of climatic change in terms of alterations in biome distribution and sea levels [8]. This subject will be discussed more extensively in Chapter 11.

Tundra

Tundra (Figs 4.7 and 4.8) is found around the Arctic Circle, north of the tree-line. Smaller areas occur in the Southern Hemisphere on sub-Antarctic islands. Alpine tundra occurs above the tree-line on high mountains, including those in the tropics. Arctic tundra is the most continuous of biomes and the easiest to define. Winter temperatures may be as low as −57°C; water melts at the soil surface in summer (air temperature is rarely over 15°C) but there is always a permanent layer of frozen soil underneath — the *permafrost*.

Alternate freezing and thawing of the soils sets up patterns of water movement in the soil that results in the development of regular surface stripes of stones, polygonal patterns (see Fig. 10.1) created by ice wedges driving into the ground, and regular earth hummocks [9]. For the vegetation there is a very short growing season, and only cold-tolerant plants can survive, as can be seen in the climate diagram in Fig. 4.8.

Typical plants are mosses, lichens, sedges and dwarf trees. Large herbivores include reindeer (caribou) and musk ox. Small herbivores include snowshoe hares, lemmings and voles. Many birds migrate there from the south in summer, feeding on the large insect populations in the tundra during that season. Carnivores are Arctic fox, wolves, hawks, falcons and owls [10].

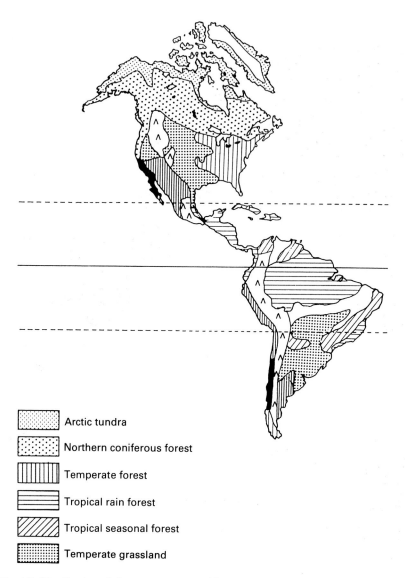

Arctic tundra

Northern coniferous forest

Temperate forest

Tropical rain forest

Tropical seasonal forest

Temperate grassland

Fig. 4.7 Distribution of the major terrestrial biomes of the world.

The impact of human beings in the tundra has been relatively slight until recent times, when the mineral resources of the high latitudes have made the region more attractive to humankind. The low productivity of the vegetation, coupled with peaty soils that are easily compressed in summer, render the tundra very sensitive to vehicular traffic, resulting in local erosion [11]. This has become particularly severe and extensive in areas of the Arctic that have been exploited for their oil reserves. Exploitation of such areas as Alaska for their fossil fuel reserves is

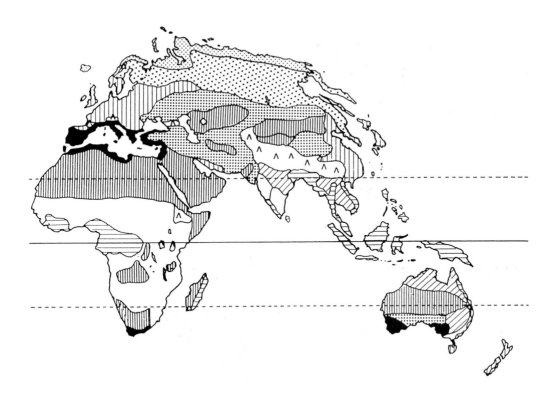

 Tropical savanna grassland and scrub

Desert

Mediterranean vegetation: chaparral

Mountains

Fig. 4.7 (Continued)

bound to continue into the future, bringing further stress to this biome, but emphasis is now being placed on landscape planning and the conservation of wildlife habitats [12]. Roads are being constructed on thick pads of gravel, up to 2 m deep, in order to prevent the melting of the underlying permafrost and the development of erosion.

Another consequence of oil exploitation in the tundra regions is the threat of oil spillage. This is mainly a problem of coastal environments, as in the case of the *Exxon Valdez* spill in 1989, when almost a quarter of a million barrels of oil were lost into the sea causing extensive damage to coastal habitats and the death of large numbers of seabirds

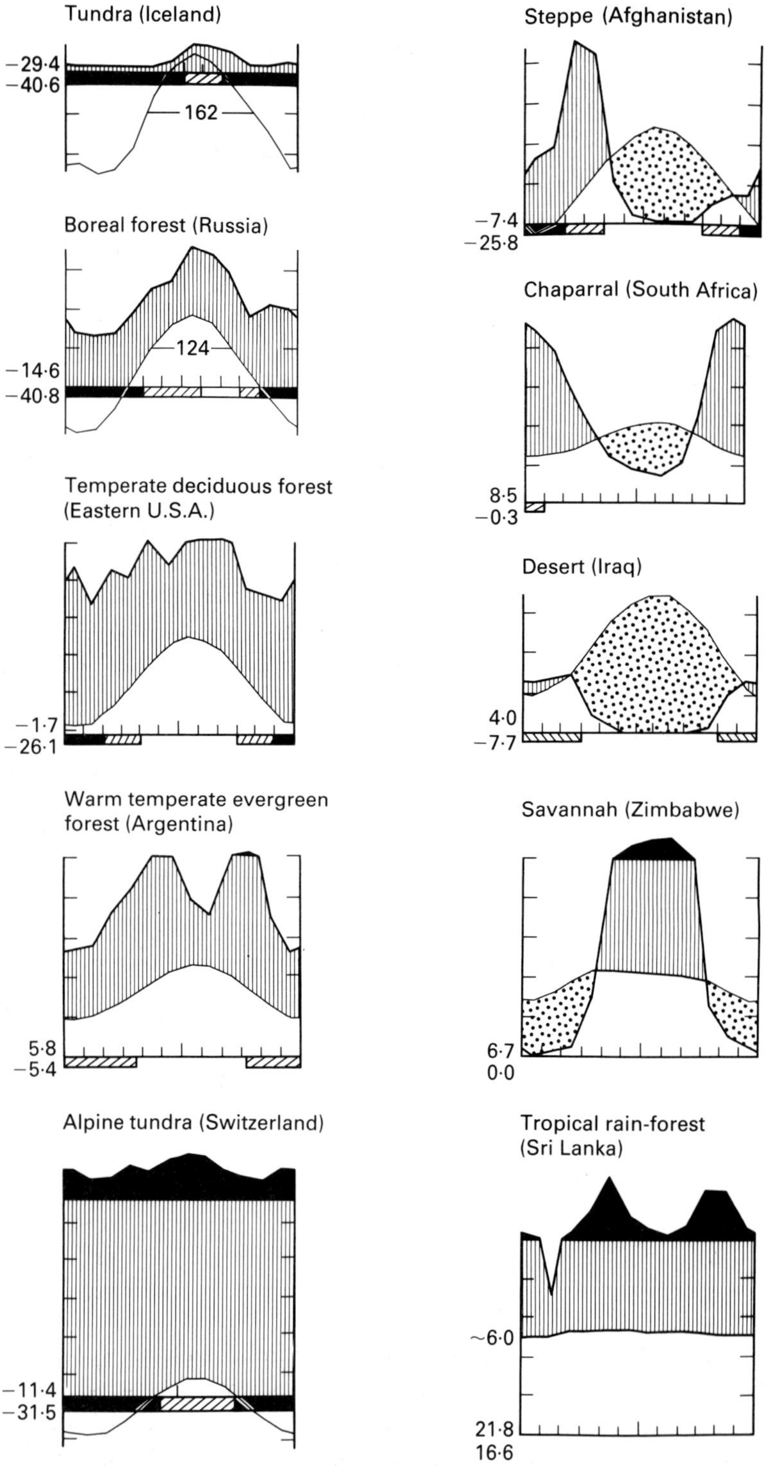

Fig. 4.8 Representative climate diagrams from the major terrestrial biomes. For key to symbols used on these, see Fig. 4.5.

[13]. Mineral extraction is a further possible future cause of disturbance of both the Arctic and Antarctic tundra ecosystems.

Another source of pollution in the tundra region comes from further afield, namely radioactivity from nuclear testing and reactor accidents. The accident at the Chernobyl nuclear power station in 1986 in what was then the Soviet Union, for example, produced radioactive caesium in dust form that accumulated in the snows over Greenland and can still be detected in the growing ice masses of glaciers [14]. The uptake of such pollutants by slow-growing lichens in tundra has had extensive repercussions, for these plants are a major food resource for reindeer (caribou), which are in turn used as food by many human cultures in tundra regions. Slow nutrient turnover ensures that the problem will remain acute for many years.

The disturbance of the tundra biome is responsible for other impacts on caribou. The construction of long pipelines to carry oil has created problems with the seasonal migration patterns of these animals, but the growing populations of humans in the tundra have also benefited caribou by reducing the density of wolves, which are their major predators and cause considerable mortality among the young animals. A balance between the two species will require some careful population management of these species [15].

The destruction of ozone in the stratosphere by chemicals produced by humankind, such as CFCs, has received considerable public attention because of the consequent development of an 'ozone hole' over the Antarctic each spring. Concern centres upon whether the hole will become larger, or appear in other regions, such as the Arctic, and also whether the hole will permit levels of ultraviolet radiation to rise sufficiently to endanger living organisms, including humankind. The oceanic phytoplankton may be particularly sensitive to such influence. Observations so far have not yielded clear results on either question. The influence of the Antarctic ozone hole does seem to extend as far as Australia and New Zealand [16], but harmful effects on living organisms have not yet been conclusively demonstrated. Although small temporary holes have been recorded in the Arctic ozone layer, it seems that the ozone in northern latitudes is generally more persistent than in the south [17].

Northern coniferous forest (taiga)

Taiga (Figs 4.7 and 4.8) forms an almost unbroken belt across the whole of northern North America and Eurasia [18]. Its northern margin with the tundra often has an intermediate zone of birch and dwarf shrubs. Its southern limit is less definite and grades into deciduous woodland. Taiga is also found on high mountains in lower latitudes, such as the southern Rockies. In the northern forests the winters are long and cold, the summers short and often very warm. The soil in winter is mostly frozen to a depth of about 2 m, but thick snow cover can keep soil temperatures as high as −7°C. Trees are mostly evergreen conifers, able to photosynthesize all year and to resist drought (a result of strong winds and extreme cold) with their needle-shaped waxy leaves. They remain undamaged by snowfalls because of their overall shape. Taiga usually contains vast tracts of one or two tree species only, except along rivers. Animals in this biome are limited by severe winters and the small number of different habitats. The most important large herbivores are deer — more species live here than in any other biome. Rodents are plentiful and can burrow under snow and survive harsh winters. Carnivores include wolves, lynxes, wolverines, weasels, mink and sable; omnivorous bears are also found. Birds either are adapted to feeding in taiga (e.g. crossbill) or are summer migrants feeding on the vast seasonal swarms of insects.

The value of timber to mankind has resulted in considerable exploitation of the coniferous forests, especially in the last two centuries. In some countries, such as Sweden, forest exploitation began very early, dating back to the seventeenth century, because of the need for fuel to extract metal from the native iron ores of that country [19]. The result of such a long history of forest depredation is that very little primary, untouched forest is left throughout Fennoscandia (Finland, Norway and Sweden). In North America, history is repeating itself and many of the remaining wilderness areas of coniferous forest, such as the red cedar forests of the west coast, are rapidly falling to commercial forestry. Even the discovery of the pharmaceutical value of some trees, such as the Pacific yew (*Taxus brevifolia*), whose bark yields a potent anti-cancer drug [20], seems to have led to intensified forest destruction rather than to the careful conservation that would logically be indicated. Siberia has almost 8 million square kilometres of boreal forest (nearly 60 per cent of the world's area of coniferous forest) and the vastness of the resource has led to very wasteful exploitation in the past. A shifting culture of nomadic logging camps forms the basis of the timber industry in that region.

The fur trade has also placed a strain on the life of the taiga, having in the past exploited both herbivores, such as beaver, and carnivores, like mink, fox and sable. Fur farming, coupled with a reducing public demand and the rise of more efficient man-made alternative fabrics, will hopefully result in the decline of trapping in the wild.

It is possible that the boreal forests are playing an important part in the regulation of the carbon dioxide content of the atmosphere, and hence the global greenhouse effect. Work by Rosanne D'Arrigo and colleagues from Columbia University, New York, suggests that some boreal forest trees in North America are growing faster as a result of rising levels of atmospheric carbon dioxide and are thus acting as a major sink for the world's excess carbon [21]. There is a strict limit to the extent to which a given area of forest can act as a sink since it attains its maximum biomass and then stabilizes, but a policy of extending rather than reducing the area currently occupied by boreal forest would help to contain the current rise in carbon dioxide.

Besides the problems resulting from clearance of boreal forest for timber, human impact in the form of air pollution may be adding to the strains placed upon this biome by our species [22]. In Europe and North America the burning of fossil fuels has resulted in the emission of waste gases, such as sulphur dioxide and oxides of nitrogen, which dissolve to form strong acids in the rainfall. These gases are often carried to regions far beyond those in which they are produced, and the boreal forest zone in both continents has proved particularly susceptible to this kind of pollution. Many of the soils of the northern regions are already acid because of the calcium-poor rocks from which they have developed, and they therefore have little capacity to neutralize acid precipitation, so that they are easily acidified by acid rain. Acid snow is of particular importance, since it accumulates through the winter and then melts to provide a sudden flush of acidity into rivers and lakes, which can prove damaging to fish stocks. The direct impact on the trees themselves can result both from damage to needles and attack by pathogens through the damaged leaf surfaces, and from the influence of increasing soil acidity on the uptake of nutrients by roots. One soil toxin in particular, aluminium, has been shown to become increasingly soluble and available to plants as soil acidity increases.

The boreal regions are rich in peatlands and these have also been extensively exploited as a source of energy-rich organic material for fuelling power stations, and also as a source of horticultural peat. These peatlands are a source of methane gas, which is an efficient greenhouse gas in the atmosphere, and any rise in global temperature may add to the rate of methane emission and thus act as a positive feedback mechanism. The cumulative impacts of humankind on the boreal forest regions have placed this biome under considerable stress in recent years.

Temperate forest

There are four basic types of temperate forest. (1) Mixed forest of conifers and broad-leaf deciduous trees. This was the original climax vegetation of much of north-central Europe, eastern Asia, and north-east North America; little remains today. (2) Mixed forests of conifers and broad-leaf evergreens. This once covered much of the Mediterranean lands but very little is left. It still occurs in the Southern Hemisphere, in Chile,

New Zealand, Tasmania and South Africa. (3) Broad-leaf forests almost entirely of deciduous trees. This formerly covered much of Europe, northern Asia and eastern North America, and is found in the Southern Hemisphere only in Patagonia. (4) The rare broad-leaf forest consisting almost entirely of evergreens. This occurs throughout much of Florida, and also in north-east Mexico and in Japan. In the Southern Hemisphere it occurs on the southern tip of South Island, New Zealand. All these regions have very high rainfall, and the dripping forests have been termed 'temperate rain forests'. In all temperate forests there is frequently an understorey of saplings, shrubs and tall herbs, which is particularly well developed near the forest edge or where human interference has occurred. Temperate forests have warm summers but cold winters, except on western seaboards. Winter temperatures may fall below freezing point. The deciduous trees escape these cold winters by losing their leaves; many plants have underground over-wintering organs. The fauna includes bears, wild boar, badgers, squirrels, woodchucks, many insectivores and rodents. Predators include wolves and wild cats (on the decline), red foxes and owls. Large herbivores are the deer. This biome is extremely rich in bird species, especially woodpeckers, titmice, thrushes, warblers and finches.

The long history of human impact on the deciduous forests of Europe has in many areas led to their destruction and replacement with agricultural land. Some remaining forest has been managed as an open canopy landscape over many centuries and this has led to the development of distinctive and often species-rich habitats. Intensive studies of ancient managed woodland in England, by Oliver Rackham of Cambridge University [23] have provided a clearer understanding of the processes involved in old management systems that have led to the development of diverse assemblages of plants and animals. This work has helped to emphasize the need to accept the role of human beings in the woodland ecosystem and the necessity for maintained management if the richness of the ancient forests in Europe is to be retained.

The climate that permits the development of temperate deciduous forest has proved particularly appropriate for the settlement of our own species, especially once the cultural developments in our history associated with agriculture, urbanization and industrialization had taken place. For this reason the biome has been very extensively fragmented, and those relict areas that remain have to cope with high local population densities of humans. Fragmentation, associated with hunting, has led to the loss of many large mammal species in Europe, such as the bison and the wolf, from most areas, and the extinction of some, such as the wild ox (aurochs). Others, such as the wild boar, are in decline.

The intensity of industrial activity within the biome has led to similar problems of air pollution to those experienced by the boreal forest. The effects of acid precipitation, for instance, were first noticed in the 1970s in the forests of southern Germany [24], where silver fir, Scots pine and beech have been particularly affected. Coniferous trees seem to be more susceptible to acid rain influence than deciduous ones [25], but damage to oak and beech has been increasing in Germany since about 1983. It is possible that a new type of damage has now become noticeable, perhaps caused by different agents, such as ozone, which is generated in the atmosphere by an interaction of vehicle exhaust gases with sunlight. Ozone damage to trees has been reported from California since the early 1950s and this pollutant gas may be the cause of increased forest damage in Europe in recent years. The possibility that ozone damage to crop plants within the temperate zone will increase, especially with anticipated climate changes in a greenhouse world, must also be carefully considered [26].

Tropical rain forest

Tropical rain forest occurs between the Tropics of Cancer and Capricorn in areas where temperatures and light intensity are always high and rainfall is greater than 200 cm a year (and is at least 12 cm in the driest month) [27]. There is a great variety of trees (see Chapter 2): in some parts of the Brazilian rain forests, there are as many as 300 species of trees in $2 \, km^2$. The popular image of the jungle — thick, steamy and impenetrable — is borne out only in those areas that man has at some time cleared, especially along river margins; true climax tropical forest has very little undergrowth. The canopy is extremely dense; the light intensity below may be as low as 1 per cent of that above, and thus only a few extremely shade-tolerant plants can survive there. Life is concentrated in the canopy, where there is plenty of light. The crowns of the trees are covered with *epiphytes* — plants that use the trees only for support and are not parasites. *Lianas* — vines rooted in the ground but with leaves and flowers in the canopy — are also characteristic. Dead plants are rapidly decomposed, so there is little undecayed plant matter on the forest floor. The rate of turnover of nutrients is very high, and the tropical forest has a higher productivity than that of any other terrestrial biome.

The soils of tropical rain forests, however, are poor in nutrients [28], in part because of the very heavy demand placed upon them by the growing plants. The epiphytes depend entirely on rainfall and the leachate from rain percolating through the canopy to satisfy their nutrient demands, and the pockets of organic soil that accumulate around them are evidently relatively rich resources for plant nutrients. This has led to the unusual phenomenon of some tree roots actually growing upwards and tapping into the nutrient stores contained within these epiphyte microhabitats [29].

The tropical rain forest biome contains the greatest variety of animal life of any biome, because of the richness of the food resources that it offers and the relative constancy of the conditions of the environment through the year. There is a great profusion of birds with many different diets — seeds, fruit, buds, nectar, or insects. Many of the mammals are adapted to arboreal life (monkeys, sloths, ant-eaters, many small carnivores) but there are also many ground-living forms, including rodents, deer and peccaries. Amphibia, and reptiles, especially snakes, are important as predators of small vertebrates and invertebrates.

The importance of structural complexity and other factors in maintaining the high diversity of both plant and animal species within tropical forests has already been discussed (Chapter 2).

The rate of loss of tropical rain forest is one of the most serious environmental problems currently facing the world. As much as 50 000 square miles (130 000 square kilometres) of rain forest are currently being lost each year and at this rate the rain forests will be entirely lost within the next 30 years. The bulk of this loss of forest is occurring in the Amazon Basin, especially in Brazil [30], but important losses are taking place throughout the tropics. Figure 4.9 shows the losses in

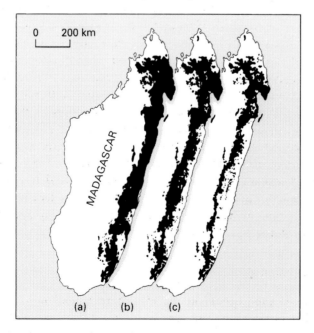

Fig. 4.9 Distribution of rain forest in the island of Madagascar: (a) the probable original extent, (b) in 1950 and (c) in 1985. Data from Green and Sussman [31]. Copyright 1990 by the AAAS.

Madagascar [31]. There are several reasons why this is a matter for concern. The process of conversion of rain forest to grassland for beef production involves considerable loss of biomass, and the carbon stored in this biomass is transferred to the atmosphere, thus adding to the greenhouse effect. The same applies to the organic matter from the soil, which is also oxidized to carbon dioxide. The loss of forest also has an influence on local climate, both by changes in the reflectivity of the surface and by the loss of the transpiration process that takes water from the soil and returns it to the atmosphere, maintaining high humidity. Since the rain forests contain about half of the world's species of plants and animals (see Chapter 2), their loss involves the extinction of a large proportion of these. Quite apart from the ethical question of whether humanity has a right to contemplate this, such a loss deprives our species of future opportunities to exploit this genetic resource for our own benefits.

It is now widely recognized that the shifting agriculture practised by indigenous populations of the rain forest, which has long been regarded as inefficient and wasteful, is actually a well-adapted and sustainable method of making a living in the forest without doing irreparable harm to the environment [32]. Even the hunting of local fauna by Amazonian Indians, which was once thought to deplete game resources rapidly and to limit the human populations, does not seem to operate in this way. Observations over a 10-year period have shown that game resources are sustainable at the levels employed by local populations [33].

Most of the countries that harbour the rain forests are economically poor, and it is rather hypocritical of developed countries to expect these nations to forgo the immediate financial gains of forest exploitation when the temperate, developed nations have either destroyed their own forests already (as in the case of the European nations), or are in the course of destroying their forests at the present (as in the case of the United States). Encouragement to manage rain forest in a sustainable way may take the form of persuasion that there are long-term financial advantages, or may be accompanied by direct financial reward in the form of debt exchange. Attempts to evaluate the sustainable yield of rain forest in economic terms are still in their infancy [34], but this is an important aspect of future research in these areas.

Temperate grassland

Temperate grassland occurs in regions where rainfall is intermediate between that of desert and of temperate forest, and where there is a fairly long dry season. Temperate grassland has many local names — the *prairies* of North America, the *steppes* of Eurasia, the *pampas* of South America, and the *veld* of South Africa — but the dominant plants in all of them are the grasses, the most widespread and successful group of land plants. The soil always contains a thick layer of humus, unlike some forest soils, but is more exposed than the latter, and therefore more likely to dry out. The dominant animals are large grazing mammals — on the North American prairies, vast herds of bison and prong-horn (which humans had virtually wiped out by the close of the last century, but are now reintroducing); over the steppes of Eurasia, the saiga antelope, wild horse and wild ass once roamed in herds; in the South American pampas the natural grazer is the guanaco; and in Australia the kangaroos fill this role. All these have been largely replaced by humans with domestic grazing animals, often with disastrous results, although grasses are adapted to withstand the effects of natural grazing.

Some of the worst problems experienced in the temperate grassland regions have been associated with over-exploitation by intensive grazing, with the conversion of unsuitable areas to arable agriculture, and with the use of fire. The Dust Bowl experience of the southern United States earlier this century is a prime example of the type of problem that can result from such misuse of this biome. Disturbance of the rather fragile temperate grassland ecosystem has often resulted in considerable changes in the species composition of the vegetation of these areas, especially when mobile human populations have provided a means of transporting species from one part of the world to another. In the sagebrush steppes of the Great Basin of Utah and Nevada, for example, a Mediterranean grass, the cheatgrass (*Bromus tectorum*) has become a serious pest that has greatly modified the composition of vegetation during this century [35]. The result of the introduction and spread of this annual grass has been that the grasslands have become more prone to fire, and this in turn has influenced the success of shrub regeneration, so that the whole character of the vegetation is now changing. The bunchgrasses, which were formerly the main grass types occupying this region, formed scattered patches that did not carry fire well, so that fire frequency, extent and intensity were lower.

It is anticipated that the climatic changes accompanying rising atmospheric carbon dioxide levels will have considerable impact on the temperate grasslands. Increasing drought and higher temperatures in these areas will lead to the extension of semi-arid and even desert conditions. This will be of particular economic importance in regions where the grasslands have been converted to arable land and have become vital grain-producing areas (see Chapter 8).

Tropical grassland or savanna

Savanna is a term applied to a range of tropical vegetation from pure grassland to woodland with much grass [36]. It covers a wide belt on either side of the Equator between the Tropics of Cancer and Capricorn. The climate is always very warm and there is a long dry season, and thus the plants often have drought-resisting features. The grass is much longer than that of temperate grassland, growing to 3·5 m. There is often a great variety of trees, which also show drought-resisting features; a typical group is the acacias. The dominant animals are large grazing mammals, the African savanna having the greatest variety, and burrowing rodents are also found. Large carnivores, such as lions and hyenas, prey on the grazers.

Fire is a natural feature of the savanna grasslands and is also employed extensively for management by the human cultures that use the savanna as grazing rangelands. In South-East Asia, for example, the dry season lasts for between five and seven months, and during that time the daily maximum temperature always exceeds 20°C, which makes the climate ideal for fire. In some areas the soil is shallow or sandy and is inadequate for the support of full forest vegetation, even in the absence of fire [37]. Fires in South-East Asia often occur between November and February, and the trees cope with this either by dropping fruit early (in the case of species that need fire for seed germination) or they may flower during the fire period and drop fruits onto the charred, but nutrient-enriched soil at the end of the fire season (as in the case of many dipterocarp trees).

Fires may be induced by lightning strikes or spontaneous combustion, but are often started by local people for a variety of reasons, such as land clearance. In this way most areas are burned each year and are relatively easily controlled because of the generally low biomass of combustible material. More serious and damaging fires are often associated with situations where a deliberate policy of fire protection has been adopted. Under these circumstances there is a build-up of dead grass and litter with the result that any accidental fire is intense and generates high temperatures. Such fires are difficult to control and can be more damaging to tree populations (compare the Yosemite National Park in 1989).

It has been suggested that the frequent burning of the tropical savannas represents a significant return of carbon back into the atmosphere, which adds to the greenhouse effect.

Savanna soils are generally poor in their nutrient content, and the decomposer system is active in returning elements from organic matter back into a mineral form. This process may limit the availability of some nutrients, like nitrogen, and some dominant grass species are known to release chemicals that inhibit the activity of nitrifying bacteria and in this way ensure that nutrients do not become available, so that more robust and competitive species are not able to assume dominance over them. Local nutrient enrichment may also be the cause of observed behaviour patterns in herds of grazing herbivores such as one finds in the tropical grasslands of East Africa. In the Seregenti National Park of Tanzania, S.J. McNaughton of Syracuse University has investigated the question of why multi-species herds of animals are frequently concentrated into particular grazing areas, and he has found that they often group in sites where nutrient availability is greater and the quality of the forage is therefore improved [38,39]. An understanding of such processes is clearly of great importance in determining management and conservation policies in these important wildlife areas.

Some of the large mammals of the savanna are under threat as a result of habitat loss because of the increase in grazing by domestic animals and because of the effects of hunting. Over the last 10 years, for example, the population of African elephants in the wild has been reduced by at least a half.

Chaparral

Chaparral occurs where there are mild wet winters and pronounced summer droughts (known as Mediterranean climate [40]), and in areas with less rain than grasslands. The vegetation is *sclerophyllous* (hard-leaf) scrub of low-growing woody plants, mainly evergreen with hard, thick, waxy leaves — adaptations to drought. In the Northern Hemisphere it occurs mainly in countries fringing the Mediterranean basin, but also in north-west Mexico and California. Formerly this biome had a varied flora and fauna, with many herbivores such as ground squirrels, deer and elk, and mountain lions and wolves as their predators, but this has been greatly reduced by humans. In the Southern Hemisphere there are areas of chaparral in southern Australia, southern Chile, and South Africa.

The Mediterranean basin itself has probably suffered from the activities of human modification of habitats for longer than any other area of the earth [41]. The outcome of millennia of over-exploitation is degraded vegetation, erosion of soils and desiccated landscapes. The extent to which this loss in productivity can be blamed upon human activity and the possible role of climate change in the process is discussed further in Chapter 11, but vegetation destruction and overgrazing have certainly played an important part. At least 25 species of plant have been lost in Israel this century, mainly as a consequence of changes in land use associated with loss of wetland habitats [42]. Other habitats are now being lost at an alarming rate as many of the hills are being planted with conifers, especially *Pinus halepensis*.

The agreeable climate of Mediterranean climate areas continues to make them attractive to human beings and many of these areas are heavily populated. Apart from the land used in urban and rural development, the problem of air pollution is becoming increasingly severe. The effect of ozone and a cocktail of other air pollutants on the coastal sage scrub vegetation of California, for example, is particularly severe, and the decline in productivity of the natural vegetation renders it less able to recover from fire [43]. This opens the community for invasion by a number of aggressive alien weeds, especially annual grasses, that are gradually assuming more important roles in the ecosystem.

Grasses from the Mediterranean Basin have proved very successful (and, from the human point of view, troublesome) invaders of almost all areas of Mediterranean climate throughout the world, including Australia and Chile [44]. Although the predominant movement of pests has been from the Mediterranean basin out to other areas, there have been notable exceptions, such as the prickly pear (*Opuntia ficus-indica*), a cactus from America that is now widespread, both in the Mediterranean countries and in South Africa. In Australia it has been a serious pest in the past. There has also been some movement of fauna between areas of Mediterranean climate at the hands of humankind. In the Mediterranean climatic area of California, for example, about 9 per cent of the breeding bird species are invaders, as well as 12 species of mammal. Some of these have been deliberately introduced, while others have escaped from captivity. Most probably have little impact on the native wildlife, but some certainly do, such as the wild pig (*Sus scrofa*), which is very destructive of vegetation. Such invasions are a matter for concern because of the very high proportion of endemic species in the Mediterranean areas, whose limited geographical distribution renders them particularly at risk of extinction.

Deserts

Areas experiencing extreme drought form deserts. A good definition is those areas where rainfall is less than 25 cm per year, or — if higher — is mostly lost immediately by evaporation. Deserts can be divided into hot deserts (such as the Sahara) with very high day-time temperatures, often over 50°C, and low night-time temperatures below 20°C with relatively mild winters, and cold deserts (such as the Gobi Desert in Mongolia) with very severe winters and long periods of extreme cold. Typical desert has large areas of barren rock or sand and very sparse vegetation. Desert plants are adapted to drought in various ways: some have drought-resistant seeds; others have small thick leaves that are shed in dry periods; yet others, such as the New World cacti, are succulents, storing water in their stems. Desert animals are mostly small enough to hide under stones or in burrows during the intense day-time heat in hot deserts. Certain rodents are well adapted to desert life — they live in cool burrows, are largely nocturnal, and waste very little water in their urine. Insects and reptiles lose little water, having waterproof skins and excreting almost dry, crystalline urine.

Deserts are currently expanding at an accelerating rate [45]. In the 70 years between 1882 and 1952 the estimated area of the land surface occupied by desert rose from 9·4 to 23·3 per cent. In the last 50 years the lands bordering the southern edge of the Sahara Desert in Africa have lost about 250 000 square miles (650 000 km^2) of grazing land. Several factors have undoubtedly contributed to this process of desertification: climatic shifts since about 1965 have certainly reduced the rainfall from monsoons penetrating into the southern Sahara (Fig. 4.10) and also into north-west India [46]. But the problems presented by such reduced precipitation have been exacerbated by the demands of local human populations for maintained food production, leading to failed crops and overgrazing. Political boundaries and population pressures often make it difficult for peoples experiencing drought and famine to escape by mass migration, and the outcome is yet further strain on the very limited resources of this biome.

Long-term climatic change and the geological history of deserts are discussed in Chapter 11. The future of the biome in a greenhouse world seems to be one of increasing aridity and continuing or accelerating desert spread [47]. Much of the western-central area of the United States is at risk, as is a large area of central Asia and Australia.

The widening of the desert belt in Africa will have biogeographical repercussions in other biomes. Many bird species from Europe, for example, cross this region twice a year in their migrations and their capacity to cross this inhospitable tract of country is delicately balanced.

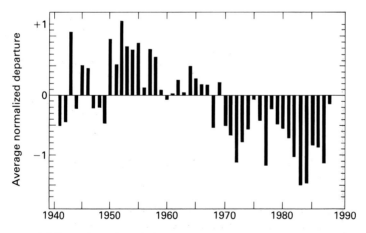

Fig. 4.10 Rainfall trends in the Sahel region of the southern Sahara expressed as departure from the long-term mean (0). Note the year-to-year fluctuations, but the general trend to low penetration of the monsoon rains since about 1970. Data from Grainger [46].

An increasingly lengthy desert crossing may well cause a rise in mortality among migratory birds, and there is some indication that this is already occurring. Figure 4.11 shows one population index for an insectivorous bird, the whitethroat (*Sylvia communis*), based on regular trappings at selected sites throughout the British Isles by the British Trust for Ornithology. This bird passes through the arid zone on its migration to West African wintering quarters, and the population index can clearly be seen to correlate with the rainfall data in Fig. 4.10, especially with the very dry years of 1983–1985 when a population crash of whitethroats was observed in the British Isles. If this is typical of such migrants, we must expect considerable changes in temperate bird life as a consequence of global climatic change.

Freshwater biomes

These biomes are far less self-contained than those of the surrounding land or the open sea. They receive a continual supply of nutrients from the land, but much of this is washed downstream in the rivers. The waters are generally far less rich in nutrients than oceans, and freshwater bodies are usually less productive than either sea or land environments. They are more changeable than ocean or land biomes; rivers gradually wear away the land through which they pass and thus the river biome

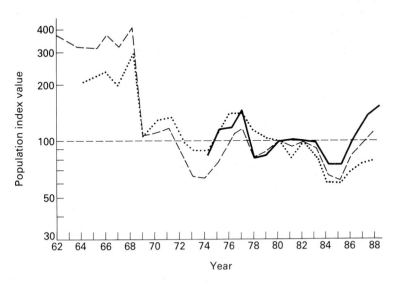

Fig. 4.11 Variations in the population density (an index is expressed on a logarithmic scale) for the whitethroat (*Sylvia communis*), an insectivorous migratory bird, in the British Isles between 1962 and 1988. The three lines represent three different censuses, but are in good agreement. Note the correspondence of population troughs with low precipitation in the Sahel (Fig. 4.10). Data from British Trust for Ornithology.

itself gradually changes, and many small ponds are seasonal, drying up in summer. There is a wide range of freshwater environments, from small ponds and streams to vast lakes and wide rivers [48]. At the lower end of the scale they are often better considered merely as a wet extension of the surrounding terrestrial biome. The dominant plants of larger lakes and slow rivers are phytoplankton, but larger floating and rooted plants cover considerable areas. Many of the animals are restricted to the freshwater habitat; amphibians, though living on land, need fresh water in which to breed; land animals use fresh water for drinking and bathing; and many birds are adapted to the freshwater habitats. Animal communities in large lakes correspond to planktonic, nektonic and benthic communities of the oceanic biome (see below). Some large lakes have well-defined shores, constituting sub-biomes: examples are the Great Lakes with their dune systems, or Lake Victoria with its muddy shores. Marshes (salt marshes and freshwater mires) are best considered as intermediate between marine or freshwater biomes and the surrounding terrestrial biome, and estuaries are transitional both between freshwater and marine biomes and also between the water and the land. They have a very complex structure and are highly productive, with a great variety of plant and animal life. Freshwater habitats have suffered greatly from pollution by man — toxic industrial wastes, detergents, and vast quantities of sewage are dumped into rivers and lakes, and cause the extinction of all but a few resistant forms of life.

Drainage and reclamation of wetlands for agriculture is resulting in loss of this habitat throughout the world. The United States alone is considered to have lost about 24 000 square miles (61 000 km^2) of wetlands during the period 1950–1970 as a result of agricultural changes and coastal development [49]. In the Nile Delta of Egypt, wetlands are being reclaimed for agriculture as fast as the delta sediments build up and extend into the Mediterranean Sea (a process which is itself slowing down because of the changes in hydrology caused by the erection of the Aswan Dam). At one time there may have been about 1·5 million square miles (4 million km^2) of peatland in the world, but at least 7 per cent of this has been lost and, with the opening up of the vast resources of Canada and Russia, the loss is likely to continue.

Acid precipitation is now a further threat to the balance of remaining wetlands, particularly temperate and boreal regions [50]. The study of diatoms (algae with persistent silica cases) in lake sediments has now clearly demonstrated that there have been significant increases in acidity in many lakes. In one study of Scottish lakes, for example, Roger Flower and Rick Battarbee [51] of the University of London showed that over the past 130 years the pH had dropped by a full unit (this is equivalent to a 10-fold increase in acidity). Acidification in Ontario lakes is proceeding so fast that it has proved possible to document the annual decrease in pH (about 0·1 pH units over five years) [52]. The recovery of some lakes following the reduction of sulphur emissions from nearby sources has proved possible [53], which raises hope for the rehabilitation of polluted areas.

Many lowland wetlands in agricultural regions are suffering from eutrophication — a form of nutrient enrichment (largely nitrate and phosphate) in which the productivity of the ecosystem first rises and then falls as the decomposer system depletes the oxygen content of the water. Oxygen depletion results in the successive loss of oxygen-demanding organisms and finally even those capable of coping with moderate levels of anoxia. The causes of eutrophication include leaching of excess fertilizer from agricultural land, excessive input of human or animal sewage, and the conversion of pastoral grasslands to arable land with the resultant mineralization of elements formerly bound up in soil organic matter.

Marine biomes

Land covers only 29 per cent of the earth's surface, whereas the oceans take up 71 per cent, with an average depth of 3900 metres. It is impossible to distinguish regional biomes in the seas, because of the uniformity of the marine environment and of the ease of distribution of its inhabitants. On the land, animals and plants of different latitudes have different life-forms; in the sea, animals do have distinctive forms, but these vary according to the depth at which they live, rather than according to latitude — for example, deep-sea animals are of a life-form especially adapted to cope with high pressures and total darkness [54]. Water has a higher specific heat than soil or rock, and even the warmest oceans never reach the high temperatures of tropical forests or hot deserts. Similarly, the coldest seas are never as cold as the tundra or northern forests. The surface temperature is never greater than 30°C and rarely falls below 0°C. Marine organisms obviously have no problems in obtaining sufficient water, but light is a limiting factor. The tiny photosynthetic plants (phytoplankton) are restricted to the upper photic zone (the uppermost 200 m); virtually no light penetrates below 500 m. Atmospheric oxygen and carbon dioxide are plentiful at the surface and these gases are also dissolved in the water. Pressure is an important factor limiting the downward movement of shallow-water species. Sea water is much richer in nutrients than fresh water, and these are recycled to the photic zone by upwellings of deep currents. In some other areas, surface waters converge and descend. Where these are already exhausted of nutrients, the area of descent forms a 'desert', such as the Sargasso Sea in the southern North Atlantic. Such areas are the only virtually unproductive parts of the surface waters.

There are three principal marine biomes. (1) The *oceanic biome* of open water, away from the immediate influence of the shore. This is further divided into the *planktonic sub-biome* containing free-floating plankton (mostly microscopic organisms with buoyancy mechanisms); the *nektonic sub-biome* of active swimmers, including fish, squids, turtles and marine mammals; and the *benthic sub-biome*, whose fauna is especially adapted for life on the sea-floor. (2) The *rocky shore biome* is dominated by large brown seaweeds. These show a distinct pattern of zonation up the shore as a response to the increasingly long period of desiccation which is experienced as one moves upwards from the low-water mark. Benthic animals display a similar zonation. (3) The *muddy* or *sandy shore biome* is constantly receiving a supply of mud or sand which accumulates and forms an unstable substrate for attachment. Few algae can survive (for example the sea lettuce, *Ulva*), but many animals, such as burrowing worms, bivalve molluscs and crustaceans, live in the sediment. These are preyed upon by wading birds.

Some wetland ecosystems fall within the intertidal zone of muddy shore habitats, including mangroves [55] and salt marshes. Not only are these important areas for many species of animals, but they also provide coastal protection for resident human populations in these regions, as in Bangladesh. Their exploitation and destruction by people is creating coastal erosion in some areas.

Of the various marine biomes, the oceanic biome is the one least influenced by human activity, though the seas of the continental shelf regions already suffer from eutrophication from rivers and sewage waste, together with the output of toxic materials from industrial areas. These toxins may be causing direct damage to living organisms and may also be weakening their resistance to diseases. The Baltic Sea, being relatively isolated from major oceanic currents, is particularly sensitive to the build-up of toxins and excessive nutrients, but even the North Sea, though more open in its circulation, receives effluent from many major rivers, including the Rhine and the Thames, and is becoming increasingly polluted.

In the tropical oceans, especially the Caribbean, there is concern over the health and functioning of highly productive coral reefs [56]. Coral consists of colonies of small animals living together with associated simple photosynthetic algae in lime-encrusted matrix that builds up into reefs. Many of these corals are becoming bleached due to the death of the algal component, and this is likely to be fatal for the whole coral system. What is causing coral bleaching is still unsure, but it could be a consequence of high water temperature, or increased ultraviolet radiation, or changes in salinity. Whether any of these factors is determined by human activity is also doubtful.

One area of marine ecology in which the human species is undoubtedly playing a major role is in the removal of fish stocks from the oceans. The increasing efficiency with which fish can be located and caught has further led to the depletion of populations of those species which are of value either as human food or even as sources of fertilizer. Careful management of fish and marine mammal resources (including whales and seals) needs to be based on thorough monitoring of numbers and control of harvests. This, of course, demands international co-operation in a biome where political boundaries do not exist beyond the coastal zone.

References

1 Ellenberg, H. (1988) *Vegetation Ecology of Central Europe*, 4th (English) edn. Cambridge University Press, Cambridge.

2 Rodwell, J.S. (1991) *British Plant Communities*, Vol. I: *Woodlands and Scrub*. Cambridge University Press, Cambridge.

3 Perkins, D.F. (1978) The distribution and transfer of energy and nutrients in the *Agrostis–Festuca* grassland ecosystem. In *Production Ecology of British Moors and Montane Grasslands* (eds O.W. Heal & D.F. Perkins). Springer Verlag, Heidelberg, pp. 375–395.

4 Oort, A.H. (1970) The energy cycle of the earth. *Scientific American* **223**(3),54–63.

5 Walter, H. (1979) *Vegetation of the Earth*, 2nd edn. Springer Verlag, Heidelberg.

6 Whittaker, R.H. (1975) *Communities and Ecosystems*, 2nd edn. Macmillan, New York.

7 Wigley, T.M.L. (1991) Could reducing fossil-fuel emissions cause global warming? *Nature* **349**,503–506.

8 Wyman, R.L. (1991) (ed.) *Global Climate Change and Life on Earth*. Chapman & Hall, London.

9 Krantz, W.B., Gleason, K.J. & Caine, N. (1988) Patterned ground. *Scientific American* **259**(6),44–50.

10 Sage, B. (1986) *The Arctic and its Wildlife*. Croom Helm, London.

11 Bliss, L.C. (1990) Arctic ecosystems: patterns of change in response to disturbance. In *The Earth in Transition* (ed. G.M. Woodwell). Cambridge University Press, Cambridge, pp. 347–366.

12 Walker, D.A., Webber, P.J., Binnian, E.F., Everett, K.R., Lederer, N.D., Nordstrand, E.A. & Walker, M.D. (1987) Cumulative impacts of oil fields on northern Alaskan landscapes. *Science* **238**,757–761.

13 Marshall, E. (1989) Valdez: the predicted oil spill. *Science* **244**,20–21.

14 Davidson, C.I., Harrington, J.R., Stephenson, M.J., Monaghan, M.C., Pudykiewicz, J. & Schell, W.R. (1987) Radioactive cesium from the Chernobyl accident in the Greenland ice sheet. *Science* **237**,633–634.

15 Bergerud, A.T. (1988) Caribou, wolves and man. *Trends in Ecology and Evolution* **3**,68–72.

16 Atkinson, R.J., Matthews, W.A., Newman, P.A. & Plumb, R.A. (1989) Evidence of the mid-latitude impact of Antarctic ozone depletion. *Nature* **340**,290–294.

17 Larsen, S.H.H. & Henriksen, T. (1990) Persistent Arctic ozone layer. *Nature* **343**,124.

18 Larsen, J.A. (1980) *The Boreal Ecosystem*. Academic Press, New York.

19 Gamlin, L. (1988) Sweden's factory forests. *New Scientist* **117**(1597),41–47.

20 Joyce, L. (1991) Scientists take wide variety of approaches to taxol studies. *The*

Scientist **5**(25),15—22(9 December).

21 D'Arrigo, R., Jacoby, G.C. & Fung, I.Y. (1987) Boreal forests and atmosphere—biosphere exchange of carbon dioxide. *Nature* **329**,321—323.

22 Mohnen, V.A. (1988) The challenge of acid rain. *Scientific American* **259**(2),14—22.

23 Rackham, O. (1980) *Ancient Woodland*. Edward Arnold, London.

24 Blank, L.W. (1985) A new type of forest decline in Germany. *Nature* **314**,311—314.

25 Schulze, E.-D. (1989) Air pollution and forest decline in a spruce (*Picea abies*) forest. *Science* **244**,776—783.

26 Ashmore, M.R. & Bell, J.N.B. (1991) The role of ozone in global change. *Annals of Botany* **67**(Supplement 1),39—48.

27 Jordan, C.F. (1982) Amazon rain forests. *American Scientist* **70**,394—401.

28 Mabberley, D.J. (1992) *Tropical Rain Forest Ecology*, 2nd edn. Chapman & Hall, New York.

29 Putz, F.E. & Holbrook, N.M. (1989) Strangler fig rooting habits and nutrient relations in the Llanos of Venezuela. *American Journal of Botany* **76**,781—788.

30 Fearnside, P.M. (1990) Deforestation in Brazilian Amazonia. In *The Earth in Transition* (ed. G.M. Woodwell). Cambridge University Press, Cambridge, pp. 211—238.

31 Green, G.M. & Susmann, R.W. (1990) Deforestation history of the eastern rain forests of Madagascar from satellite images. *Science* **248**,212—215.

32 Eden, M.J. (1987) Traditional shifting cultivation and the tropical forest system. *Trends in Ecology and Evolution* **2**(11),340—343.

33 Vickers, W.T. (1988) Game depletion hypothesis of Amazonian adaptation: data from a native community. *Science* **239**,1521—1522.

34 Peters, C.M., Gentry, A.H. & Mendelsohn, R.O. (1989) Valuation of an Amazonian rainforest. *Nature* **339**,655—656.

35 Billings, W.D. (1990) *Bromus tectorum*, a biotic cause of ecosystem impoverishment in the Great Basin. In *The Earth in Transition* (ed. G.M. Woodwell). Cambridge University Press, Cambridge, pp. 301—322.

36 Stott, P. (1991) Recent trends in the ecology and management of the world's savanna formations. *Progress in Physical Geography* **15**,18—28.

37 Stott, P. (1988) Savanna forest and seasonal fire in South East Asia. *Plants Today* **1**,196—200.

38 McNaughton, S.J. (1988) Mineral nutrition and spatial concentrations of African ungulates. *Nature* **334**,343—345.

39 McNaughton, S.J. (1990) Mineral nutrition and seasonal movements of African migratory ungulates. *Nature* **345**,613—615.

40 Suc, J.P. (1984) Origin and evolution of the Mediterranean vegetation and climate in Europe. *Nature* **307**,429—432.

41 Attenborough, D. (1987) *The First Eden: The Mediterranean World and Man*. Collins, London.

42 Naveh, Z. & Kutiel, P. (1990) Changes in the Mediterranean vegetation of Isreal in response to human habitation and land use. In *The Earth in Transition* (ed. G.M. Woodwell). Cambridge University Press, Cambridge, pp. 259—299.

43 Westman, W.E. (1990) Detecting early signs of regional air pollution injury to coastal sage scrub. In *The Earth in Transition* (ed. G.M. Woodwell). Cambridge University Press, Cambridge, pp. 323—345.

44 Groves, R.H. & Di Castri, F. (1991) *Biogeography of Mediterranean Invasions*. Cambridge University Press, Cambridge.

45 Spooner, B. & Mann, H.S. (1982) *Desertification and Development: Dryland Ecology in Social Perspective*. Academic Press, London.

46 Grainger, A. (1990) *The Threatening Desert: Controlling Desertification*. Earthscan Publications, London.

47 Petit-Maire, N. (1990) Will greenhouse green the Sahara? *Episodes* **13**(2),103–107.
48 Harper, D. (1992) *Eutrophication of Freshwaters*. Chapman & Hall, London.
49 Williams, M. (1990) (ed.) *Wetlands: A Threatened Landscape*. Basil Blackwell, Oxford.
50 Schindler, D.W. (1988) Effects of acid rain on freshwater ecosystems. *Science* **239**,149–157.
51 Flower, R.J. & Battarbee, R.W. (1983) Diatom evidence for recent acidification of two Scottish lochs. *Nature* **305**,130–133.
52 Dillon, P.J., Reid, R.A. & De Grosbois, E. (1987) The rate of acidification of aquatic ecosystems in Ontario, Canada. *Nature* **329**,45–48.
53 Gunn, J.M. & Keller, W. (1990) Biological recovery of an acid lake after reductions in industrial emissions of sulphur. *Nature* **345**,431–434.
54 Gage, J.D. & Tyler, P.A. (1991) *Deep Sea Biology: A Natural History of the Organisms at the Deep-sea Floor*. Cambridge University Press, Cambridge.
55 Rodriguez, G. (1987) Structure and production in Neotropical mangroves. *Trends in Ecology and Evolution* **2**,264–267.
56 Roberts, L. (1987) Coral bleaching threatens Atlantic reefs. *Science* **238**, 1228–1229.

5 *The source of novelty*

Some insects are protected from detection by predators by having an almost perfect resemblance to a leaf or a twig. This is perhaps the most dramatic example of the intricate way in which an organism is adapted to its environment. Other adaptations are just as intricate and thorough, although not so obvious. Every aspect of the environment makes its demand upon the structure or the physiology of the organism: the average state of the physical conditions, together with their daily and annual ranges of variation; the changing patterns of supply and abundance of food; the occasional increased losses due to disease, to predators, or to the increased competition from other organisms at the same level in the ecological food web. Every species of animal or plant must be adapted to all these conditions; it must be able to tolerate and survive the hostile aspects of its environment, and yet able to take advantage of its opportunities.

In the nineteenth century it was accepted that each species that we see today had always existed precisely as we now see it. God was thought to have created each one, with all its detailed adaptations, and these had remained unchanged. Fossils were considered to be merely the remains of other types of animal, each equally unchanging during its span of existence, which God had destroyed in a catastrophe (or a number of catastrophes) such as the biblical Flood.

In his journey round the world in the ship HMS *Beagle* from 1831 to 1836, Darwin saw two phenomena that eventually led him to consider alternative explanations. On the Galápagos Islands in the Pacific, isolated from South America by 600 miles (960 km) of sea, different birds were well adapted to feeding on different diets. Some, with heavy beaks, cracked open nuts or seeds; some, with smaller beaks, fed on fruit and flowers; others again, with fine, narrow beaks, fed on insects. On the mainland, these different niches are occupied by quite different, unrelated types of bird — for example, by toucans, parrots and flycatchers. The remarkable fact was that on the Galápagos Islands each of these varied niches was instead filled by a differently adapted species of one type of bird, the finch. It looked very much as though finches had managed to colonize the Galápagos Islands before other types of bird and then, free from their competition, had been able to adapt to diets and ways of life that were normally not available to them. This logical explanation, however, ran directly against the current idea of the fixity of characteristics. Equally disturbing were the fossils that Darwin had found in South America. The sloth, armadillo and guanaco (the wild ancestor of the domesticated llama) were each represented by fossils that were larger than the living forms, but were clearly very similar to them. Again, the idea that the living species were descended from the fossil species was a straightforward explanation, but one that contradicted the view that each species was a special creation and had no blood relationship with any other species.

Natural selection

The explanation that Darwin eventually deduced and published in 1858 is now an almost universally accepted part of the basic philosophy of biological science. Darwin realized that any pair of animals or plants produces far more offspring than would be needed simply to replace that pair: there must, therefore, be competition for survival amongst the offspring. Furthermore, these offspring are not identical with one another, but vary slightly in their characteristics. Inevitably, some of these variations will prove to be better fitted for the mode of life of the organism than others. The offspring that have these favourable characteristics will then have a natural advantage in the competition of life, and will tend to survive at the expense of their less fortunate relatives. By their survival, and eventual mating, this process of *natural selection* will lead to the persistence of these favourable characteristics into the next generation.

Evolution is therefore possible because of competition between individuals that differ slightly from one another. But why should these differences exist, and why should each species not be able to evolve a single, perfect answer to the demands that the environment makes upon it? All the flowers of a particular species of plant would then, for example, be of exactly the same colour, and every sparrow would have a beak of precisely the same size and shape. Such a simple solution is not possible, because the demands of the environment are neither stable nor uniform. Conditions vary from place to place, from day to day, from season to season. No single type can be the best possible adaptation to all these varying conditions. Instead, one particular size of beak might be the best for the winter diet of a sparrow, while another might be better adapted to its summer food. Since, during the lifetimes of two sparrows differing in this way, each type of beak is slightly better adapted at one time and slightly worse adapted at another, natural selection will not favour one at the expense of the other. Both types will therefore continue to exist in the population as a whole, as in the case of the oystercatchers described on p. 64.

Because we do not normally examine sparrows very closely, we are not aware of the many ways in which the individual birds may differ from one another. In reality, of course, they vary in as many ways as do different individual human beings. In our own species we are accustomed to the multitude of trivial variations that make each individual recognizably unique — the precise shape and size of the nose, ears, eyes, chin, mouth, teeth, the colour of the eyes and hair and the type of complexion, the texture and waviness of the hair, the height and build, the pitch of voice, and degree of resistance to different stresses and diseases. We know of other, less obvious characteristics in which individuals also differ, such as their fingerprints and their blood group. All of these variations are, then, the material upon which natural selection can act. In each generation, those individuals that carry the greatest number of less advantageous characteristics would be least likely to live long enough to have children who would perpetuate these traits, while those with a large number of more advantageous characteristics would be more likely to survive and breed successfully.

Changes of this kind in the characteristics of a species are not merely theoretical deductions, but can be shown to have taken place. This can be seen clearly only when the environment of a species has changed rapidly, which does happen, though rarely. A particularly clear example resulted from the darkening of the countryside around the industrial cities of Great Britain during the second half of the nineteenth century. This change greatly affected those moths, such as the peppered moth, *Biston betularia*, which relied upon camouflage to protect them from being seen and eaten by insectivorous birds. As long as the bark of the trees on which they rested was pale, it was advantageous for the moths to be pale also. But as industrialization proceeded and the bark of trees near the cities became blackened by soot, the pale individuals of *B. betularia* were now more and more conspicuous (Fig. 5.1). It was no coincidence that it was in 1848 that a dark or 'melanic' form of this moth first appeared and gradually became more and more common in these industrial areas until, by 1895, it was the pale form that was now the rare exception near the cities. The reason for this change is quite clear. Against the soot-darkened bark it was now the melanic form that was less conspicuous and therefore favoured by natural selection [1]. Experiments have confirmed this deduction. A large number of peppered moths, some light and some dark, each marked by a tiny spot of paint, were released in two areas. In the first, a non-industrial area, later trapping led to the recapture of 14·6 per cent of the pale coloured moths, but of only 4·7 per cent of the melanic forms — far more of these had already been eaten by birds. In the industrial area the proportions were reversed: only 14 per cent of the pale moths were recaptured, but 27·5 per cent of the dark forms, which here had been better camouflaged than their paler relatives. It is also interesting to find that, near towns that have in recent years taken measures to reduce smoke production, the proportion of melanic moths has already dropped.

Fig. 5.1 (a) Photograph showing the inconspicuousness of the normal form and the conspicuous appearance of the melanic form of the peppered moth (*Biston betularia*) on an unpolluted, lichen-covered tree trunk in Dorset, England. (b) The reverse situation when the same two forms are on a soot-covered trunk near Birmingham, England. From the experiments of Bernard Kettlewell, University of Oxford [1].

The peppered moth is not the only known example of a rapid change in the characteristics of an insect due to the influence of man. The use of DDT to control insects conferred a great advantage upon those that were resistant to this chemical, and a high proportion of house flies are now of the DDT-resistant variety. The evolution of strains of bacteria resistant to commonly used antibiotics provides other clear examples of Darwin's principle of natural selection in action.

An interesting botanical parallel to this zoological response to industrial pollution is provided by the bent grasses *Agrostis stolonifera* and *A. tenuis* living around a copper refinery that was established in northern England in about 1900 [2]. The soil from uncontaminated grassland in the area contained less than 100 parts per million of copper; this grassland contained many species of plant, and samples of the two *Agrostis* species showed very little tolerance to higher levels of copper. At the other extreme, old-established lawns near the refinery were composed exclusively of *A. stolonifera* and *A. tenuis*, and these were living in soils containing 2600–4200 p.p.m. of copper. Interestingly enough, some lawns that had been established in the last 10 years, in which the soil contains

1900—4800 p.p.m. of copper, bore a greater variety of species of grass, and the species of *Agrostis* showed a lower mean level of tolerance. This suggests that the process of selection of copper-tolerant forms is still taking place in these small populations.

If all the members of a species gradually came to possess such new characters as a resistance to DDT, the species would have changed but no additional species would have resulted. However, as Darwin realized in the Galápagos Islands, the original single species may also split into two or more new species. In order to explain how this happens, the meaning of the term 'species' must first be explained. Why do biologists consider that a sparrow and a robin are separate species, but that dogs such as a German shepherd and a greyhound (which appear just as different from one another) are both members of the same species? To biologists the essential difference is that, under normal conditions in the wild, a sparrow and a robin do not mate together, while a German shepherd and a greyhound will (the great difference in the appearance of the two dogs is due to artificial selection by man). Sometimes the difference is a little more subtle, as in the case of the horse and the ass; though these are separate species, they do sometimes breed together. However, this cross between the two species is short-lived and does not result in a permanent merging, because the resulting mule or hinny is sterile.

There are, then, two phenomena that have to be explained: the splitting of one species into two, and the inability of these two species to breed together. As will be seen, these are really two aspects of a single problem, and a clue has already been provided by the melanic form of the peppered moth. This is the common variety around the industrial cities of Britain, where experiments showed it was better adapted than the pale variety, while the reverse is true in country districts far from the cities. Within the species as a whole, more than one type of adaptation is now found, but the distribution of each is distinct. The process of adaptation has taken place independently in each population, not in a general way throughout the species.

This independent adaptation of each population is possible because each is, to some extent at least, isolated from other populations. Though the distribution of the peppered moth covers the whole of the mainland of Great Britain, in fact the moth is found almost exclusively within patches of light woodland. It is only found in the intervening areas of open country if high winds have blown it from its normal woodland habitat. As a result, the peppered moths of each patch of woodland will, over long periods of time, mate only with one another. This, then, is the source of the isolation that permits the independent adaptation of each population.

Each species is broken up in this way into separate populations, cut off by areas in which physical or biological factors make it difficult or impossible for the species to survive. As long as the isolation persists, each population will gradually tend to become slightly different from the others, due to the action of two forces. One of these forces lies within the animal or plant, the other lies outside it.

The isolating force within the organism

The force within the organism lies in the system that is responsible for the transmission of the characters of the parents to the next generation. Within each cell lies a rather opaque object called the *nucleus*, inside which is a number of thread-like bodies called *chromosomes*. These chromosomes consist of a chain of large complex molecules known as *genes*. It is the biochemical action of these genes that is responsible for the characteristics of every cell of an individual, and thus for the characteristics of the organism as a whole. There might, then, be a particular gene that determined the colour of an individual's hair, while another might be responsible for the texture of the hair and another for its waviness. Each gene exists in a number of slightly different versions, or *alleles*. Taking the gene responsible for hair colour as an example, one allele might cause the hair to be brown while another might cause it to be red. Many different alleles of each gene may exist, and this is the main reason for much of the variation in structure that Darwin noted.

An individual of course inherits characteristics from both of its parents. This is because each cell carries not one set of these gene-bearing chromosomes, but two: one set derived from the individual's mother, and the other derived from its father. A double dose of each gene is therefore present, one inherited from the mother, the other from the father. Both parents may possess exactly the same allele of a particular gene. For example, both may have the allele for brown hair, in which case their offspring would also have brown hair. But very often they may hand down different alleles to their offspring; for example, one might provide a brown hair allele, while the other provided a red hair allele. In such a case the result is *not* a mixing or blurring of the action of the two alleles to produce an intermediate such as reddish-brown hair. Instead, only one of the two alleles goes into action, and the other appears to remain inert. The active allele is known as the *dominant* allele and the inert one as the *recessive* allele. Which allele is dominant and which recessive is normally firmly fixed and unvarying — in the hypothetical example given, the brown hair allele might be dominant, and the red hair allele might be recessive.

The genes themselves are highly complex in their biochemical structure. Though normally each is precisely and accurately duplicated each time a cell divides, it is not surprising that from time to time — due to the incredible complexity of the molecules involved — there is a slight error in this process. This may happen in the cell divisions that lead to the production of the sexual gametes (the male sperm or pollen, the female ovum or egg). If so, the individual resulting from that sexual union may show a completely new character, unlike those of either of its parents. In the example given above, such an individual might have completely colourless hair. Such sudden alterations in the genes are known as *mutations*.

The genetic system outlined above can lead to changes in the characteristics of an isolated population in two ways. Firstly, new mutations may appear and spread through the population. Secondly, since each individual carries several thousand genes, and each may be present in any one of its several different alleles, no two individuals carry exactly the same genetic constitution, or *genotype* — unless they are identical twins, developed from the splitting of a single original developing egg. Inevitably, therefore, the two isolated populations will differ somewhat in their initial genetic content, some alleles being rarer in one population than in the other or, in extreme cases, being absent altogether. As mating goes on in the two populations, new combinations of alleles will appear haphazardly in each, and this will lead to further differences between them.

Whether they are new mutations, or merely new recombinations of existing alleles, new characteristics will therefore appear within an isolated population. Any of these which confer an advantage on the organism are likely to spread gradually through the population, and so change its genetic constitution. However, it is important to realize that chance, as well as its genetic constitution, plays a role in determining whether a particular individual survives and breeds. Even if a new, favourable genetic change appears in a particular individual, it may by chance die before it reproduces, or all its offspring may similarly die, so that the new mutation or recombination disappears again. But, however rare each genetic change may be, each is likely to reappear in a certain percentage of the population as a whole. In a larger population, each mutation or recombination will therefore reappear sufficiently often that the effects of random chance are nullified, and the underlying advantages or disadvantages that they confer will eventually show themselves as increased or decreased reproductive success. For this reason, it is the population, and not the individual, that is the real unit of evolutionary change. In smaller populations, however, chance will play a greater role in controlling whether a particular allele becomes common or rare or disappears; this effect is known as 'genetic drift', because it is not directed by selective pressures.

The way in which the genotype is expressed, as the morphology, physiology, behaviour, etc., of the organism, is known as the *phenotype*. This is somewhat variable and can be modified by the environment.

Thus identical twins (sharing therefore an identical genotype) will come to differ from one another if they are brought up in areas with, for example, differing amounts of sunlight or of available food. This slight plasticity of the genotype is valuable from an evolutionary point of view, for it makes it possible for a single genotype to survive in slightly different habitats.

External isolating forces

It is, then, the independent appearance of new mutations in each population, and the independent course of genetic change within populations, that together make up the driving force within the organism that tends to make each isolated population gradually become different from every other. The force *outside* the organism that aids the process is simpler. Natural selection acts to adapt the population to its surroundings. But no two patches of woodland, no two freshwater ponds, will be absolutely identical, even if they lie in the same area of country. They may differ in the precise nature of their soil or water, in their range of temperature, or their average temperature, or in the particular species of animal or plant that may become unusually rare or unusually common in that locality. Since each population has to adapt to slightly different conditions, the two populations will gradually come to differ from one another.

The history of a patch of sunflowers living in a ditch in the Sacramento Valley of California provides a good example of the way in which all these forces can gradually make two populations become quite different from one another [3]. The population consisted of natural hybrids between the annual sunflowers of California, *Helianthus annuus* and *H. bolanderi*. To begin with, the original population gradually became split into two by a drying-out of part of the ditch, the dry section being colonized by grasses among which the sunflowers could not survive. Over the space of five years the dry grassy patch widened until the two sub-populations of sunflowers were over 100 metres apart. One of these was now in a deeper part of the ditch, which remained wet until late spring, while the other grew in a shallower, drier position. The two sub-populations became different in a number of characteristics, such as the shape of the flower head as a whole, the number of sterile floret rays surrounding the head, the shape of the base of the leaf, and the length of the hairs on the stem and leaves. Even though bees could easily fly from one population to the other, so that some cross-pollination between them must have taken place, observations over the next seven years showed that the differences between the two populations did not disappear. Their environments differed sufficiently to preserve the distinctiveness of the two populations, which varied in size from nearly 5000 to nearly 75 000 plants, depending on the amount of spring rainfall, but was never small enough for random genetic drift to have an effect. The hybrids between the two species not only survived throughout this period of time, but also formed the majority of the population in the drier, eastern part of the ditch. The fertility of the pollen of these hybrids increased from c. 10 per cent in the beginning to over 80 per cent by 1955. This suggests that the hybrids were well adapted to their environment and that, consequently, natural selection was favouring those individuals. In the wetter, more western population, most of the plants looked like *H. bolanderi*, and a relatively small number looked like *H. annuus*; but only a few looked intermediate between the two species. So these two populations showed an increasing evolutionary divergence over the space of only a few years.

A similar example of the effects of a change in environment is shown by the experiment of two British biologists, Malhotra and Thorpe. In 1991 they translocated populations of the lizard *Anolis oculatus* from different areas of the island of Dominica in the West Indies, which differed considerably in their altitude, climate and vegetation, to large enclosures in the north-west of the island [4]. The mortality rates of these populations over the next two months showed a selective disappearance of certain phenotypes from these populations, and the intensity of this was significantly correlated with the degree of change in ecological conditions that each population had experienced.

Once populations have started to diverge in their genetic adaptations in this way, the foundations for the appearance of a new species have been laid. If two divergent populations should meet again when the process has not gone very far, they may completely hybridize and merge into one another. The further, vital step towards the appearance of a new species is when hybrids between the two independent populations do appear, but only along a narrow zone where the two populations meet. Such a situation suggests that, though continued interbreeding within this zone can produce a population of hybrids, these hybrids cannot compete elsewhere with either of the pure parent populations. This seems to be the situation with the woodpecker-like flickers of North America. The eastern yellow-shafted flicker, *Colaptes auratus*, does mate with the western red-shafted flicker, *C. cafer*, along a narrow 2000-mile-long (3200 km) stretch where the two meet, but this zone of hybridization does not seem to be spreading (Fig. 5.2).

There can be no general rule as to the length of time that it will take for the descendants of one original species to diverge so far from one another in their genetic constitution that they have become separate species. The most important factor in determining the rate of genetic change is the speed at which the environment changes, for this determines the extent to which the organism is no longer well adapted to its

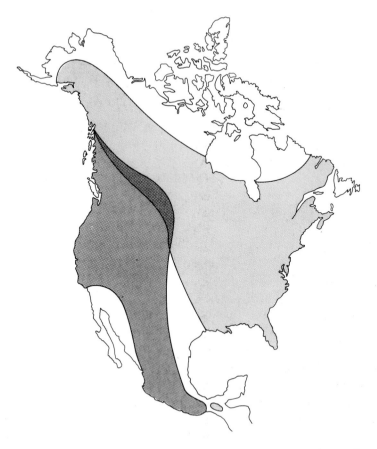

Fig. 5.2 The yellow-shafted flicker of eastern and north-western North America hybridizes with the red-shafted flicker of the south-west over a long but narrow zone.

environment. If the environment changes rapidly, the organism must also change rapidly, or else become liable to extinction. For example, if the peppered moth or the bent-grasses described earlier had been unable respectively to change their colour or their tolerance of copper, they would have become extinct. But the rate at which an organism can respond to environmental changes is also dependent on population size. In a small population the random effect of genetic drift may by chance produce a new mixture of genetic characteristics that match the new requirements of the environment. This is less likely to happen in a larger population, where the sheer size of the gene pool makes rapid evolutionary change of this kind less likely.

The fastest well-documented example of speciation is that of some species of cichlid fish in a small lake in Africa. This became separated from the large Lake Victoria by a strip of land that has been dated by radiocarbon analysis at 4000 years old [5], and the species of fish in the small lake therefore cannot be older than that. Many of the species of the fruit-fly *Drosophila* that are found in the narrow valleys between the lava flows on the slopes of the volcanic mountains in the Hawaiian Islands (see p. 157) are probably also only a few thousand years old.

Barriers to interbreeding

Once independent evolution in isolation has produced a situation in which the hybrids are less well-adapted than either of their parents, then natural selection will favour individuals that do not perpetuate this more poorly adapted hybrid population. This may be either because they cannot, or will not, mate with individuals from the other group, or because such unions are infertile. The barrier to hybridization is known as an *isolating mechanism*, and it may take many forms. In animals such as birds and insects, that have a complicated courtship and mating behaviour, small differences in these rituals may in themselves effectively prevent interbreeding. Sometimes the preference for the mating site may differ slightly. For example, the North American toads *Bufo fowleri* and *B. americanus* live in the same areas, but breed in different places [6]. *B. fowleri* breeds in large, still bodies of water such as ponds, large rainpools and quiet streams, whereas *B. americanus* prefers shallow puddles or brook pools. Interbreeding between species is also hindered by the fact that *B. americanus* breeds in early spring and *B. fowleri* in the late spring — though there is some mid-spring overlap.

Many flowering plants are pollinated by animals that are attracted to the flowers by their nectar or pollen. Hybridization may then be prevented by the adaptation of the flowers to different pollinators. For example, differences in size, shape and colour of the flowers of related species of the North American beard-tongue (*Pentstemon*) adapt them to pollination by different insects — or, in one case, by a hummingbird (Fig. 5.3). In other plants, related species have come to differ in the time at which they shed their pollen, thus making hybridization impossible. Even if pollen of another species does reach the stigma of a flower, in many cases it is unable even to form a pollen tube, because the biochemical environment in which it finds itself is too alien. It cannot, therefore, grow down to fertilize the ovum. Similarly, in many animals alien spermatozoa cause an allergy reaction in the walls of the female genital passage and the spermatozoa subsequently die before fertilization.

Other isolating mechanisms may not prevent mating and fertilization taking place, but instead ensure that the union is sterile. These may be genetic isolating mechanisms, the structure and arrangement of the genes on the chromosomes being so different that the normal processes of chromosome splitting and pairing that accompany cell division are disrupted. These differences may make themselves felt at any stage from the time of fertilization of the ovum, through all the steps in

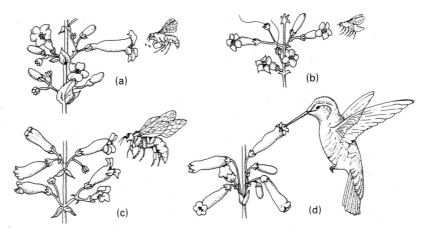

Fig. 5.3 Four species of the beard-tongue (*Pentstemon*) found in California, together with their pollinators. Species (a) and (b) are pollinated by solitary wasps, species (c) by carpenter bees, and species (d) by humming birds. After Stebbins [7].

development, to the time at which the sexual gametes of the hybrid itself are produced. Whenever the effects are felt, the result is the same; the hybrid mating is sterile or, if offspring are produced, these are themselves sterile (as in the case mentioned earlier of mating between a horse and an ass), or of reduced fertility.

Polyploids

Another method by which new species can appear is by *polyploidy* — the doubling of the whole set of chromosomes in the nucleus of a developing egg or seed, so that each automatically has an identical partner. This may occur in the development of a hybrid individual (in which case it can overcome any genetic isolating mechanisms), or in the development of an otherwise normal offspring of parents from a single species. In either case the new polyploid individual will be unlikely to find another similar individual with which to mate, and the origin of new species by polyploidy has therefore been important only in groups in which self-fertilization is common. Only a few animal groups fall into this category (e.g. turbellarians, lumbricid earthworms, and weevils), but in these groups an appreciable proportion of the species probably arose in this way. In plants, however, in which self-fertilization is common, polyploidy is an important mechanism of speciation [7]. More than one-third of all plant species have probably arisen in this way, including many valuable crop plants such as wheat (see p. 291), oats, cotton, potatoes, bananas, coffee and sugar cane. Polyploid species are often larger than the original parent type, and also more hardy and vigorous — many weeds are polyploids.

An example of a pest species resulting from such polyploidy is the cord-grass, a robust, rhizomatous plant of coastal mud flats around the world. There are several species of this plant, but none of them were serious pests until two species were brought into contact with one another in the waters around Southampton, a major port on the south coast of Britain, in the latter half of the last century. An American species of cord-grass, *Spartina alterniflora*, was brought into the area, probably carried in mud on a boat, and it was able to hybridize with the native English species, *S. maritima*. The hybrid was first found in the area in 1870, and was named *Spartina × townsendii*. It contained 62 chromosomes in its nucleus but, because these chromosomes were derived from two different parent species, they were unable to join together in compatible pairs before gamete formation, and so the hybrid did not produce fertile pollen grains or egg cells. It was nevertheless able to reproduce vegetatively, and it is still found along the coasts of western Europe. But in 1892 a new fertile cord-grass appeared near Southampton, and was named *Spartina anglica*. This has 124 chromosomes, as a result of a doubling of the number found in the sterile hybrid, so that the chromosomes could once again form compatible pairs and fertile gametes could be produced. This new species, formed by natural polyploidy, has been extremely successful and has spread around the world, often creating problems for shipping by forming mats of vegetation within which sediments are deposited, contributing to the silting up of estuaries.

Polyploidy can also be artificially produced, for example by the use of colchicum, an extract of the meadow saffron plant, *Colchicum autumnale*. Techniques of this kind have been used to provide new strains of commercially valuable plant, such as cereals, sugar-beet, tomatoes and roses.

Evolution — a summary

Whatever may be the nature of the isolating mechanism that keeps them separate, two groups have become two independent species as soon as they are no longer able to interbreed or to produce fertile hybrids. Where only one species existed before, there are now two. They may have become sufficiently different in their adaptations to be able to spread into one another's area and coexist without competing. For example, in North America two closely related species of bird, the song sparrow *Melospiza melodia* and the Lincoln sparrow *M. lincolni* live together, as do the red maple tree *Acer rubrum* and the sugar maple tree. *A. saccharum*.

The whole process of *speciation* (as the evolutionary process leading to new species is termed) is, then, able to start only because, since each organism can only exist under a limited range of conditions, and the conditions in the environment vary in an irregular fashion, each species becomes broken up into separate populations. Within each of these, new features are continually appearing due to genetic changes, and natural selection is constantly weeding out those new features least suited to the environment [8].

Adaptations for survival

The complex biochemicals of an animal or plant cell are quite unlike most inorganic substances, and their complicated interactions can only take place within a limited range of physical conditions. In order to survive, therefore, any cell must continually ensure that its biochemicals remain isolated from those of its surroundings, and that it remains within the range of conditions in which its own biochemicals can continue to function. Evolutionary history has been the gradual process by which organisms have become able to isolate themselves (or, more precisely, their body tissues) from their surroundings with increasing effectiveness. This has made it possible for organisms to become able to survive in conditions that are more and more unfavourable for life — the conquest of dry land being a major step in that direction.

Alongside the evolution of adaptations providing insulation from the physical environment, organisms have also had to cope with difficulties due to their companion species. Both the existence of animals or plants that are similar to one another in their adaptations, and which therefore compete with one another, and the complicated interactions of herbivore and plant food or of predator and prey, lead to the appearance of new difficulties. Evolution is, therefore, the process by which organisms have conquered two types of barrier: those imposed by physical conditions, and those resulting from the biological world of animals and plants among which they live.

Even in the most favourable habitats, the physical conditions are rarely ideal for the organism through the 24-hour daily cycle. In addition to the daily alternation of light and dark, with the accompanying rise and fall in temperature and relative humidity, the temperature and rainfall may vary considerably from one day to another in many areas. Any organism must be able to tolerate changes of this kind in the physical conditions: animals may take shelter during rainstorms, while plants may close their leaves or flowers.

As long as these conditions are short-lived, evasion or toleration is not difficult. A more serious problem exists in parts of the world where the climate is seasonal. Here the conditions may not be severe in themselves, but their continuation over a period of months demands a quite different adaptation on the part of the organism. Such prolonged alterations in physical conditions inevitably affect the whole community: the lower temperatures and shorter daylight hours of winter directly affect both plants and animals and, in addition, most animals find that food is then far less plentiful. Some animals, of course, are able to avoid these conditions altogether, by migrating to warmer climates. This solution is particularly common in flying animals, such as birds and butterflies, that can cover relatively long distances with ease, and to which a river or a stretch of sea is not an impassable obstacle. Other animals, such as bears and many smaller mammals, endure the cold and the scarcity of food by *hibernating* — reducing their metabolism to a minimum and surviving on food reserves they have stored up in the body during the summer. Resting stages, of one kind or another, are common in both animals and plants of higher latitudes. The hard, resistant seed-cases of many plants, which will not germinate until they have been exposed to the coldness of winter (during which the parent plant may die), have their counterpart in the periods of arrested development of many insects. The hard, resistant chrysalis of a butterfly is a stage during which the complex changes from the caterpillar to the winged adult are carried out. But this inert, non-feeding stage is equally a convenient form in which, by a slowing-down of the rate of these changes, the whole winter can be passed in comparative safety.

For plants, the dry season of areas closer to the Equator brings the risk of desiccation due to lack of water, which may also be unavailable during winter in cold-temperate latitudes because it is frozen into ice. Winter also brings the danger of frost damage. Though the stem of the plant can be protected by bark, the leaves are still exposed. The flowering plants were able to solve this problem by developing the mechanism of leaf-fall, so that the enormous, exposed leaf surface is shed completely until the following spring. It is interesting to find that this adaptation was probably first developed by flowering plants in the tropical regions to reduce water loss during the dry season [9]. Only later did flowering plants with this *deciduous* habit spread to the colder regions where water is also in short supply because it is frozen. In addition to cutting down the rate of water loss, the reduction in the exposed surface area, due to the shedding of leaves in winter, reduces the damage caused by high winds and by settling snow in these regions.

Meeting the challenge of the environment

The adaptations considered so far are all ones that in one way or another *evaded* the challenge of inhospitable conditions. To live and carry on all its normal functions in an area that is particularly cold, or particularly hot and dry, requires a more thoroughgoing adaptation of the whole organization of the organism and of its life history. For example, most frogs and toads cannot survive in desert regions because the adults quickly become desiccated and because the water that the embryos need for their development rapidly becomes too hot for their existence and eventually evaporates completely. Nevertheless, some frogs and toads have been able to adapt to these conditions. An example is the spadefoot toad *Scaphiopus couchii* which is found in the deserts of the south-western United States [10]. Its eggs are laid in temporary desert rainpools resulting from local storms, and their rate of development is very high. As a result, they pass through the most temperature-sensitive stages of their development before the afternoon of the next day, when the temperature rises above the level (about 34°C) that is critical for them. The larvae also hatch at an early stage from the mass of jelly which surrounds them, and are therefore soon able to seek the coolest parts of the rainpool. The adults can survive in these deserts because their hind limbs are modified to form scoop-like spades, with which they can excavate holes. In the hottest periods they can therefore retreat from the desert surface to the cooler, moister environment of their hole.

Even the spadefoot toad is, in retreating to its hole, still only temporarily able to evade the problems posed by the physical conditions of its environment; it will eventually have to emerge to feed. The limitations of this type of solution are as obvious as are the advantages of more fundamental adaptations that permanently insulate the organism from unfavourable physical conditions. The insulating coat, formed by hair in mammals, or by feathers in birds, helps them to maintain the internal temperature of their body at a constant level even if, like a camel or an ostrich, they live in a desert where the day-time temperature may be as high as 55°C. The insulating coat also reduces their rate of water loss to a tolerable level, as does the resistant external skeleton of insects. Similarly, the thick cuticle of the leaves of evergreen plants and of conifer needles protects them from winter conditions which other flowering plants can only survive by shedding their leaves. Thus, the evergreens are able to take advantage of warm, sunny spells early in the year for their photosynthesis, while deciduous species cannot manufacture food during the winter months when they have no leaves, and must develop a new canopy each spring.

Adaptations of this kind do not merely allow their owners to colonize regions having extreme climates. Since their internal conditions are more constant, the rates of their biochemical processes and the level of activity of these organisms can remain constant, irrespective of daily or seasonal climatic changes. Their comparative insulation from the effects of physical conditions therefore provides them with a considerable advantage in their competition with other organisms. It is no coincidence that the groups that have this insulation — the insects, birds and mammals — are by far the most numerous, varied and widespread of terrestrial animals.

Clines and 'rules'

Environmental conditions usually vary in complex fashion, so that it is difficult to distinguish any regular changes that may result from them. There are, however, exceptions to this. Such conditions as average temperature or rainfall may change gradually and regularly according to either latitude or altitude. Some aspects of the organism, such as its size or height, may then also vary gradually and continuously across the area; this regular change is known as a *cline*. For example, the yarrow *Achillea lanulosa* grows in North America from the Pacific coast to the 4000 m crest of the Sierra Nevada Mountains. The higher the altitude at which it is found, the lower is the average height of the plant. This is a genetically controlled adaptive feature, for seeds from specimens from different altitudes retain their relative height characteristics even if grown together.

Some similar changes have been noted in warm-blooded animals, and have been called 'rules'. For example, warm-blooded animals lose both heat and moisture through their body surface. Because a larger animal has a smaller surface area compared with its volume than does a smaller animal, warm-blooded animals tend to be larger in cooler, drier environments than in hotter, more humid environments, thus conserving heat and moisture; this is known as 'Bergmann's rule' [11]. Similarly, 'Allen's rule' notes that such projecting parts of the body as the ears and tail tend to be shorter in colder, drier regions, for the same reasons. However, although these 'rules' are of some historic interest, there are many other characteristics, such as brood size or frequency, which vary in this way and which are no less important. It must also be stressed that these 'rules' apply primarily to variations within a species: the differences between species are more complex, so that simple effects of this kind are often concealed by other results of their differing ways of life.

Competition for life

However successful their adaptations to their physical environment, organisms must also adapt to the demands of the biological world around them, either to avoid being eaten, or to compete for space or food supply with other organisms. There can be no final solution to any of these problems for, as quickly as new adaptations appear that reduce predation or allow more successful competition with other species, the predator or competitor will in its turn adapt. The herbivorous group that becomes able to run faster and escape from its predators, itself provides the stimulus that leads to the evolution of faster predators. The plant that evolves spines or unpleasant-tasting biochemicals to avoid being eaten by herbivores similarly stimulates the appearance of herbivores insensitive to these defences. For example, milkweeds are rich in poisonous glycosides and are avoided by most caterpillars. The caterpillars of the monarch butterfly, however, not only feed on these plants, but also in turn use the biochemical to make the adult butterfly unpalatable to birds. Other biochemicals commonly used by plants as feeding deterrents include alkaloids, flavoids, quinones and raphides (crystals of calcium oxalate) [12].

One method by which the problem of competition can be at least reduced is for the two competing groups gradually to become specialized to different ways of life; they may then be able to exist together in the same area without competing with one another.

Finally, a part of the adaptation of any population is to ensure that its numbers are approximately adjusted to the food supply of the area. The territorial behaviour of some birds, such as the Scottish red grouse (*Lagopus lagopus scoticus*), does this very effectively [13]. Each male takes possession of an area of heather moor large enough to provide an adequate food supply for its family, and defends it against other members of its species. In a year when food is scarce, the territory claimed is larger. The males compete for these territories by display, and this system therefore not only ensures that it is the weakest birds that are excluded from the moor (and are frequently killed by predators or starvation), but also ensures that an adequate food supply is available for the successful birds. This type of social competition is a close parallel to that in human societies. In both, as a result of social competition those which are successful receive a variety of advantages — sexual, nutritive and environmental. The red grouse society no longer contains a group of moderately successful males, sometimes adequately fed and at other times weakened by malnutrition. Instead, it is permanently divided into the 'haves', assured of the necessities of life and of the opportunity, by reproduction, to transmit their characteristics to the next generation, and the 'have-nots' of whom about 60 per cent die during the winter.

Controversies and evolutionary theory

Though there is now a vast amount of evidence for the theory of evolution by natural selection, controversy still exists about some details of the circumstances in which new species evolve or the rate at which this happens. For example, some biologists believe that evolutionary change normally takes place at a steady, gradual rate. Others instead believe that, even if genetic alterations gradually accumulate within a population, this may not be reflected in detectable morphological or physiological changes until they are so numerous as to shift the balance of the whole genotype. At this point a comparatively large number of changes are seen to take place at the same time; this is known as the 'punctuated equilibrium' model of evolutionary change. Each group of theorists provide examples that may support their view — and, at times, the same example is interpreted by each as supporting their own view.

It is also difficult to isolate such underlying patterns from the more direct effects of the environment. For example, a study of the fossil shells of gastropod molluscs from Pliocene and Pleistocene deposits (see Fig. 7.2) in northern Kenya shows long periods during which their structure and size remained unchanged, interrupted by shorter periods (5000 to 50 000 years) during which they changed rapidly (Fig. 5.4). This was interpreted as an example of punctuated equilibrium [15]. However, the fact that the periods of change took place in several lineages at about the same time suggests that they were the result of external events that affected all of them, rather than resulting from some inherent evolutionary mechanism.

The main difficulty with such studies is that, in general, the fossil record is not sufficiently detailed for us to be able to be certain whether gradualistic evolution or punctuated evolution was involved. In any case, we have no reason to believe that either style of evolution systematically prevails over the other. The real point of interest should instead therefore be to identify the circumstances under which one or the other type of evolution would be more likely to take place.

Another argument that concerns evolutionary biologists is whether new species always arise in isolation, separated from the area in which the ancestral, related species is to be found. This is known as 'allopatric' speciation, and is the situation most frequently found — as, for example, in the case of the separate islands within which the distinct species of finch recognized by Darwin had evolved. But some biologists believe that a new species can also arise 'sympatrically', within the area of distribution of the ancestral species. For example, two different species of lacewing insect are found in the cold-temperate and boreal regions of North America. *Chrysopa carnea* is found in grasslands, meadows and on deciduous trees, but only rarely on conifer trees; it is light green in colour. *Chrysopa downesi* is found only on conifer trees, and is a very dark green. Apart from their different colours, the two species are nearly indistinguishable morphologically and, though they will interbreed in the laboratory, they do not do so in nature. It was originally suggested [16] that the common ancestor of these two living species lived in both habitats. A simple genetic mutation could then have led to the appearance of the dark green coloration in some individuals. This coloration gave a

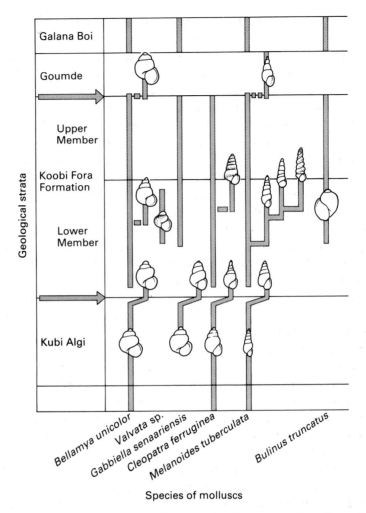

Fig. 5.4 Evolutionary changes in fossil gastropod molluscs in northern Kenya. The arrows indicate the levels at which sudden evolutionary changes took place simultaneously in several different species. From Dowdeswell [14].

selective advantage to these individuals in the coniferous habitat, but a corresponding selective disadvantage in the grassland–meadow–deciduous tree habitat. These selective forces would then have led to the appearance of two populations, each living in only one of these habitats. The consequent genetic isolation of these populations would have permitted independent evolutionary change to take place in each. It has subsequently been shown [17] that the two species also differ in their complex mating calls, and it has been suggested that this, rather than their coloration, was the original basis for the separation of the two species. However, it is equally possible that this reproductive difference evolved subsequently as a barrier to hybridization between the two populations as they became closely adapted to the two different habitats. The difficulty is that there is no evidence of these past genetic events, and it is also possible that *C. downesi* evolved allopatrically in an area in which the meadow, habitat had disappeared, and then spread back into this part of the environment in areas that were still inhabited by the light green type of *C. carnea*, which it then displaced from the conifer habitat. It is possible that the two types of oystercatcher described in Chapter 3 (p. 64), with differing thicknesses of bill, are undergoing sympatric speciation.

These controversies, and others like them, are common in any area of science, as new observations provoke new theories or suggest modifications of existing theories. But the protagonists in these disputes are only arguing about details of the theory of evolution: all accept that the theory itself is correct, and is indeed the only one that makes sense of the phenomena of the living world. It is particularly important to realize this, as some groups within society, such as the 'Creation Science' organization, are opposed to the idea of evolution and try to present these academic controversies as symptoms of widespread and fundamental scepticism of the validity of the theory of evolution. They are, of course, nothing of the kind.

Evolution and the human race(s)

As one travels around our planet, it is obvious that the types of men and women that have long inhabited the different continents differ from one another in systematic ways. They differ most obviously in physical characteristics of the skin, eyes, hair and build, but also in less obvious ways, such as aspects of their physiology and biochemistry. Should they, therefore, be regarded as separate species? The answer to that is clear. As we have seen, the distinguishing feature of members of the same species is that they can interbreed with one another without any marked reduction in fertility of the succeeding generations. Because there is no evidence whatever of any decrease in fertility resulting from matings from even the most geographically distant or apparently different types of mankind, there can be no doubt that they are all merely different races of one species, *Homo sapiens*. In any case, these racial differences account for very little of the genetic diversity of our species. Nearly all of that diversity (85 per cent) is made up of the genetic differences between the individuals within a single population (such as the peoples of Spain or a single African tribe). Another 8 per cent is made up of genetic differences between populations of this kind, and only 7 per cent of all the diversity is contributed by differences between one race and another.

Like any other species, ours has been subject to natural selection, and the racial differences are an aspect of this process. Each race has its own particular assemblage of many alleles. To some extent the process by which the alleles that come to dominate or be common in each race is haphazard, for the selective value of some of them is still uncertain. For example, though the blood group Rh− is found in 30 per cent of all Caucasians and is rare in the peoples of the Far East and in American Indians, and the AB blood group is totally absent from American Indians, there is still no clear correlation between these aspects of blood immunology and any selective advantage or disadvantage. It is known that some of the blood groups are associated with a higher or lower incidence of a particular disease. For example, individuals with blood group O are more likely to suffer from stomach ulcers, and those with group A are more likely to develop stomach cancer. But these illnesses normally only appear late in life, so that they are unlikely to have had a significant selective effect.

Despite the uncertainty over the selective reasons for their different blood group characteristics, the advantages of many of the other differences between the races are quite clear. For example, skin coloration is probably related to the fact that vitamin D (which is needed for calcium fixation and bone growth) is produced in the skin by a reaction between ultraviolet light and various precursor biochemicals. The light skin of northern Caucasians probably evolved in order to allow more of the weak northern sunlight to penetrate the skin, so as to produce enough vitamin D, while the very dark skin of more tropical races may have prevented an over-production of the vitamin, which could have been toxic. The tightly curled hair of African races helps to prevent excessive evaporation, and heat stroke, in the tropical sun. The narrower eye slit of the Asian race may help to protect their eyes from grit and glare in the deserts, and their padding of fat may help them to retain heat during the cold nights and winters. The smaller nostrils and longer noses of peoples that live at higher altitudes may help to warm and humidify the air before it passes down into the lungs.

But natural selection has not only affected such physical characteristics. A good example is the well-known history of the human blood condition known as sickle-cell anaemia, which causes anaemia and malfunctions of the liver and blood system. The gene for this condition was common in West Africa, probably occurring in over 20 per cent of the population, because it also provided resistance to malaria. In West Africans who were transported as slaves to North America, where malaria does not occur, this compensating advantage of the sickle-cell anaemia gene was no longer relevant, and the frequency of the gene has now dropped to below 5 per cent. In Central America, where malaria still persists, the gene is still found in 20 per cent of the population.

Another example of adaptive genetic change in our species is provided by the distribution of the ability of adults to digest milk. Most adults cannot do this, because they lack the enzyme lactase, which is responsible for the biochemical utilization of the lactose sugar found in milk; if they drink milk they suffer from nausea, vomiting and abdominal pains. It is interesting to find that lactose is retained into adult life in just those groups (the Europeans and some Northern African tribes) that are, or have been, pastoral nomads and for whom the milk of domesticated animals was therefore a readily available additional source of nourishment. The possession of the gene that led to the retention of lactose in adults must have provided a high selective advantage, and this has led to the frequency of the gene rising from perhaps 0·001 per cent some 9000 years ago, when the domestication of livestock commenced, to its present 75 per cent in these populations [18].

Though, via natural selection, the environment controlled the evolution of our own species until comparatively recently, we have now largely turned the tables on Nature. At first we merely modified our immediate environment to make it more congenial, by the use of fire, tools, clothing and dwellings. But now we have gone further, to modify the animals, plants and total environment of much of the planet to provide the food and energy sources we require. The human population of the earth has risen so dramatically over the last 2000 years that the impact of natural selection in determining which individual will survive to reproduce, and which will not, has been greatly diminished. To a large extent we now control the environment, rather than being selected by it. Understanding the problems that arise from this fact is the subject of much of this book.

References

1 Kettlewell, H.B.D. (1961) The phenomenon of industrial melanism in Lepidoptera. *Annual Review of Entomology* **6**,245–262.
2 Lin Wu & Bradshaw, A.D. (1972) Aerial pollution and the rapid evolution of copper tolerance. *Nature* **238**,167–169.
3 Stebbins, G.L. & Daly, K. (1961) Changes in the variation pattern of a hybrid population of *Helianthus* over an eight year period. *Evolution* **15**,60–61.
4 Malhotra, A. & Thorpe, R.S. (1991) Experimental detection of rapid evolutionary response in natural lizard populations. *Nature* **353**,347–348.
5 Fryer, G. & Iles, T.D. (1972) *The Cichlid Fishes of the Great Lakes of Africa; their Biology and Evolution*. Oliver & Boyd, Edinburgh.
6 Blair, A.P. (1942) Isolating mechanisms in a complex of four species of toads. *Biological Symposium* **6**,235–249.
7 Stebbins, G.L. (1950) *Variation and Evolution in Plants*. Columbia University Press, New York.
8 Mayr, E. (1970) *Populations, Species and Evolution*. Oxford University Press, London.
9 Axelrod, D.I. (1966) Origin of deciduous and evergreen habits in temperate forests. *Evolution* **20**,1–15.

10 Zweifel, R.G. (1968) Reproductive biology of anurans of the arid South-West, with emphasis on adaptation of embryos to temperature. *Bulletin of the American Museum of Natural History* **140**,1–64.

11 James, F.C. (1970) Geographic size variations in birds and its relationship to climate. *Ecology* **51**,365–390.

12 Ehrlich, P.R. & Raven, P.H. (1964) Butterflies and plants: a study in evolution. *Evolution* **18**,586–508.

13 Watson, A. (1977) Population limitation and the adaptive value of territorial behaviour in the Scottish red grouse, *Lagopus l. scoticus*. In *Evolutionary Ecology* (eds B. Stebbins & C. Perrins). Macmillan, London, pp. 19–26.

14 Dowdeswell, W.H. (1984) *Evolution: a Modern Synthesis*. Heinemann, London.

15 Williamson, P.G. (1981) Palaeontological documentation of speciation in Cenozoic molluscs from Turkana Basin. *Nature* **293**,437–443.

16 Tauber, C.A. & Tauber, M.J. (1977) A genetic model for sympatric speciation through habitat diversification and seasonal isolation. *Nature* **268**,702–705.

17 Henry, T.S. (1985) Sibling species, call differences, and speciation in green lacewings (Neuroptera: Chrysopidae: *Chrysoperla*). *Evolution* **39**,965–984.

18 Bodmer, W.F. & Cavalli-Sforza, L.L. (1976) *Genetics, Evolution and Man*. Freeman, San Francisco.

6 *Islands and oceans*

Isolation is one of the key factors in permitting evolutionary change, for it allows the gene pool of a population to become different from that of other populations. On large land masses, that isolation is variable in its nature and therefore in its effects on the biota as a whole. Islands provide clearer examples of isolation, for the sea surrounding them is an environment in which few terrestrial or freshwater organisms can survive for any length of time. Special adaptations for transport by air or water are necessary for an organism to cross a stretch of ocean. Dispersal to islands is therefore by a sweepstakes route, the successful organisms sharing adaptations for crossing the intervening region rather than for living within it (see p. 32). This greatly restricts the diversity of life that is capable of emigrating to an island. But, as we shall see, many factors control precisely how many organisms will reach and colonize an island, and also control the degree of diversity that its biota will ultimately attain. By comparing the biotas of islands of different size, or islands lying in different distances from their source of colonists, or islands of different topography or lying at different latitudes, we can learn much about the interaction of these factors in the control of organic diversity.

Islands are therefore of interest in three ways. Firstly, it is interesting to a biologist to observe the nature of the island biota: how it differs from that of its source-area, and the nature of the adaptations of the immigrants that allowed them to reach and to colonize the island. Secondly, it is interesting to attempt to identify and quantify the factors that control three phenomena: the rate at which new species reach an island, the rate at which species become extinct on an island, and the number of species that an island can support. Thirdly, it is interesting to study the processes of evolutionary change by which the island biota becomes an integrated ecosystem, each organism adapted to the physical and biological aspects of its life on the island, while some groups diversify to occupy ecological niches that on the mainland are normally occupied by other groups. After discussing these aspects of island biology and giving some examples of each, their interaction is examined by discussing the biota of the Hawaiian Islands.

Problems of access

Oceans are the most effective barrier to the distribution of all land animals except those that can fly. Some flying animals, such as larger birds and bats, may be capable of reaching even the most distant islands unaided, using their own powers of flight, especially if, like water birds, they are able to alight on the surface of the water to rest without becoming waterlogged. Smaller birds and bats and, especially, flying insects may reach islands by being carried passively on high winds. These animals may, in their turn, carry the eggs and resting stages of other animals, as well as the fruits, seeds and spores of plants.

Most land animals cannot survive in sea water for long enough to cross oceans and reach a distant island, but it seems possible that some may occasionally make the journey on masses of drifting débris. Natural rafts of this kind are washed down the rivers in tropical regions after heavy storms, and entire trees may also float for considerable distances. For example, in 1969 a floating island of vegetation, 13 m across and containing 10−15 trees between 6 m and 12 m tall, drifted at least 100 miles from eastern Cuba in 11 days. Such a floating island could carry small animals such as frogs, lizards or rats, the resistant eggs of other animals, and specimens of plants not adapted to oceanic dispersal. It may seem unlikely that an animal could be transported in this way and arrive safely. However, even if the odds that this happens in any one year are as low as one in a million then, over the 65 million years of the Cenozoic Era, successful colonization will almost certainly take place.

It is potentially far easier for a plant to adapt to long-distance dispersal. Very many plants show some adaptation to ensure that the next generation is carried away from the immediate vicinity of the parent [1]. It requires little elaboration of some of these dispersal devices to make it possible for them to traverse even wide stretches of ocean. In addition to this, successful colonization requires only a single fertile spore or seed, whereas in most animals it requires the dispersal of either a pregnant female or a breeding pair. The spores of most ferns and lower plants are so small (0·01−0·1 mm) that they are readily carried considerable distances by winds. Some plants have seeds that are specially adapted to being carried by the wind. Orchid seeds, for example, are surrounded by light, empty cells, and some have been known to travel over 200 km. *Liriodendron* and maple seeds have wings, and the seeds of many members of the Compositae (daisies and their relatives) have tufts of fluffy hairs; those of thistles have been carried by the wind for 145 km. Many fruits and seeds have special sticky secretions or hooks to make them adhere to the bodies of animals. Examples are the spiny fruits of burdocks and beggarticks, and the berries of mistletoe, which are filled with a sticky juice so that the seeds that they contain stick to birds' beaks. The seeds of some other plants (e.g. *Convolvulus, Malva, Rhus*) can germinate after up to two weeks in a bird's stomach.

A few plants have developed fruits and seeds that can be carried unharmed in the sea. For example, the coconut fruit can survive prolonged immersion, and the coconut palm (*Cocos nucifera*) is widespread on the edges of tropical beaches. But, since the beach is as far as most sea-borne fruits or seeds are likely to get, only species that can live on the beach are able to colonize distant islands in this way. The fruits or seeds of plants that live inland would be less likely to reach the sea and, even if they were able to survive prolonged immersion and were later cast up on a beach and germinated, they would be unable to live in a beach environment.

An example of the relative importance of these different methods of plant dispersal is provided by the Galápagos Islands [2]. Nearly all of the *indigenous* flora of these islands (i.e. those not introduced by man) is derived from that of South America, and about 378 colonizations were probably involved. Birds probably brought 60 per cent of these, 31 per cent were wind-dispersed, and 9 per cent drifted in by sea.

Variety of island habitats

Islands smaller than about nine hectares are effectively no more than beaches, because they are incapable of holding fresh water, and the flora is therefore restricted to species that are salt tolerant [3]. This leads to a corresponding reduction in the variety of animal life. The inhospitability of the beach is merely an extreme example of the problem that faces any organism, even after it has succeeded in reaching an island. This is the problem of finding a habitat in which it can survive. It is obvious that a large island is more likely to contain a greater variety of habitats than a smaller one, and that it will therefore be able to support a greater variety of forms of life. For example, studies of islands off the coast of California and Baja California have shown that their area is the most important single factor in determining the diversity of their flowering plant populations [4]. Though the diversity of available habitats also increases in islands that include higher ground, providing a cooler, moister climate, these studies showed that this factor appeared to have comparatively little effect. The effects of latitude were, however, detectable.

The very great importance of the diversity of habitat available can be shown by comparing the bat faunas of a number of islands off the northern coast of South America. Aruba, Curaçao and Bonaire, which are arid, have few species, and only one more is found in Margarita, which has little rain forest. The numbers increase considerably in Grenada, Tobago and Trinidad, where there are great areas of rain forest, and especially in Trinidad, which also has some mountains.

Problems of isolation

In this last example, getting to the islands was not a great problem, since bats can readily fly from island to island. In most cases, however, the biota is strongly affected by the degree of isolation of the island. However diverse the habitats that it offers, the variety of the island life depends, in the short term, very much upon the rate at which colonizing animals and plants arrive. This, in turn, depends largely upon how far the island is from the source of its colonizers, and upon the richness of that source. If the source is close, and if its biota is rich, then the island in its turn will have a richer biota than another, similar island which is more isolated or which depends upon a source with a more restricted variety of animals and plants. Each sea barrier further reduces the biota of the next island, which in turn becomes a poorer source for the next.

For example, the data provided by Van Balgooy [5] make it possible to map the diversity of conifer and flowering plant genera in the Pacific island groups (Fig. 6.1). This clearly shows that diversity is much lower in the more isolated island groups of the central and eastern Pacific.

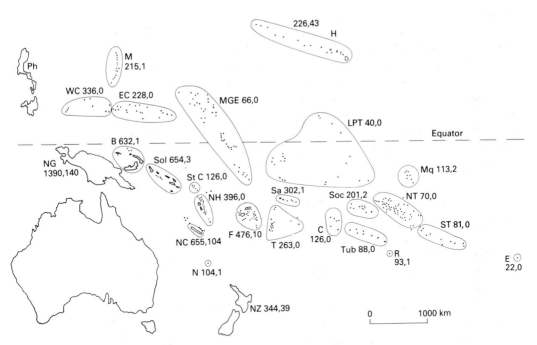

Fig. 6.1 The distribution of conifers and flowering plants in the Pacific Islands. The first number beside each island group is the total number of genera found there; the second is the number of endemic genera found there. B, Bismarck Archipelago; C, Cook Islands; E, Easter Island; EC, East Carolines; F, Fiji Islands; H, Hawaiian Islands; LPT, Line, Phoenix and Tokelau Island groups: M, Marianas; MGE, Marshall, Gilbert and Ellis Islands; Mq, Marquesas; N, Norfolk Island; NC, New Caledonia; NG, New Guinea; NH, New Hebrides; NT, Northern Tuamotu Islands; NZ, New Zealand; Ph, Philippines; R, Rapa Island; Sa, Samoa group; Soc, Society Islands; Sol, Solomon Islands; ST, Southern Tuamotu Islands; StC, Santa Cruz Islands; T, Tonga group; Tub, Tubai group; WC, West Carolines. Data from Van Balgooy [5].

However, in several of the more westerly island groups the diversity is much higher than their geographical position alone would lead one to predict. A logarithmic graph of the relationship between the number of genera and the area of the islands (Fig. 6.2) clearly shows that, in most cases, the generic diversity is simply dependent on island area. (The fact that, as a result, some islands have more genera than do the islands from which most of their flora is derived suggests that other genera were once also present in these latter islands, but have since become extinct there.) Nearly all of the more isolated island groups (shown as triangles in Fig. 6.2) have, as would be expected, a much lower diversity than would be predicted from their areas alone.

The number of land and freshwater bird species in each island shows a similar relationship to island area (Table 6.1), but this is probably due, not to island area directly, but instead to the resulting higher floral diversity.

Hazards of island life

Like any other population, the island population of a species must be able to survive periodic variations in its environment. But island life is

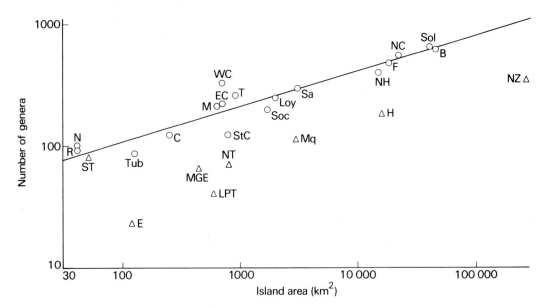

Fig. 6.2 The relationship between island area and diversity of conifer and flowering plant genera in the Pacific Islands. The more isolated islands are indicated by triangles. The data from the other islands lie very close to a straight line (the regression coefficient), suggesting that generic diversity in these islands is almost wholly controlled by island area — the correlation coefficient is 0·94, indicating a very high degree of correlation. For abbreviations, see legend of Fig. 6.1, plus Loy, Loyalty Islands. Data from Van Balgooy [5].

Table 6.1 The relationships between island area and the diversity of bird genera and non-endemic flowering plant genera in some Pacific islands. Data from Van Balgooy [5]; Mayr, [6]; MacArthur & Wilson [7]

	Area (km^2)	Angiosperm genera	Bird genera
Solomon Islands	40 000	654	126
New Caledonia	22 000	655	64
Fiji Islands	18 500	476	54
New Hebrides	15 000	396	59
Samoa group	3100	302	33
Society Islands	1700	201	17
Tonga group	1000	263	18
Cook Islands	250	126	10

more hazardous than that on the mainland, for several reasons. Catastrophe, such as volcanic eruption, has longer-lasting effects in an island situation, for there is little opportunity for a species to vacate the area and return subsequently, nor is re-invasion easy should extinction take place. On the mainland, on the other hand, chance extinction of a species in a particular area can soon be made good by immigration from elsewhere. An island will therefore contain a smaller number of species than a similar mainland area of similar ecology. For example, study of a two-hectare plot of moist forest on the mainland of Panama showed that it contained 56 species of bird, while a similar plot of shrub contained 58 species. The offshore Puercos Island, 70 hectares in area and intermediate between the two mainland plots in its ecology, contained only 20 of these species [8].

Since its success and survival are the only measures of an organism's degree of adaptation to its environment, the fact that a species has become extinct also demonstrates that it was not able to adapt to the biotic or climatic stresses to which it was exposed. The adaptation to an island environment is an unusually difficult one for a species to make. In the first place, the immigrant individuals were originally part of the mainland population, and were therefore adapted to the mainland environment. They cannot, therefore, already be adapted to the different conditions of the island. Secondly, if the colonists are few in number, they can include only a very small part of the genetic variation that provided the mainland population with the flexibility to cope with environmental change; this is sometimes known as the 'founder principle'. Finally, small populations are also far more susceptible to random non-adaptive changes in their genetic make-up. Since it is less likely to be closely adapted to its environment, a small population is also more liable to chance extinction.

A species which can make use of a wide variety of food is therefore at an advantage on an island, for its maximum possible population size will be greater than that of a species with more restricted food preferences. The advantage of this will be especially great in a small island, in which the possible population sizes are in any case smaller. This is probably the reason why, for example, though on the larger islands of the Galápagos group both the medium-sized finch *Geospiza fortis* and the small *G. fuliginosa* can coexist, on some of the smaller islands of the group there is only a single form of intermediate size [9].

Chance extinction is also a particular danger for the predators in the fauna, since their numbers must always be far lower than those of the species upon which they prey. As a result, island faunas tend to be unbalanced in their composition, containing fewer varieties of predator than a similar mainland area. This in turn reinforces the fundamental lack of variety of the animal and plant life of an island, which is due to the hazards involved in entry and colonization. The complex interactions of continental communities containing a rich and varied fauna and flora act as a buffer that can cope with occasional fluctuations in the density of different species, and even with temporary local extinction of a species. This resilience is lacking in the simple island community, and so the chance extinction of one species may have serious effects and lead to the extinction of other species. All these factors increase the rate at which island species may become extinct.

There are clearly several different possible reasons why a particular organism may be absent from a particular island. It may be unable to reach it; it may reach it but be unable to colonize it; it may have colonized it but later have become extinct, or it may simply as yet, by chance, not have reached the island [10]. It is often very difficult to decide which of these possible reasons was the cause in any particular case. In some instances, as for example where two species have complementary distributions in the same island group but are never found on the same island, the facts obviously suggest that they compete with one another so strongly that they cannot coexist.

The Theory of Island Biogeography

The number of species found on an island, therefore, depends on a number of factors — not only its area and topography, its diversity of habitats, its accessibility from the source of its colonists, and the richness of that source, but also the equilibrium between the rate of colonization by new species and the rate of extinction of existing species. Many individual observations and analyses of such phenomena have been made over the past 150 years. However, scientists always attempt to synthesize a mass of isolated data of this kind into a unifying theory that will not only explain them, but will also enable predictions to be made. Such a quantitative theory was eventually produced by the American ecologists Robert MacArthur and Edward Wilson in their book, *The Theory of Island Biogeography* [11], in which this approach by means of mathematical modelling is clearly explained.

Initially, they simply show how the position of the equilibrium point depends on the balance between the rates of colonization and of extinction (Fig. 6.3). The rate of colonization will be high initially, because the island will be reached quickly by those species that are adept at dispersal, and because these will all be new to the island. As time passes, more and more immigrants will belong to species that have already colonized the island, so that the rate of appearance of new species will drop. The rate of extinction, on the other hand, will rise. This is partly because, since every species runs the risk of extinction, the more that have arrived, the more species there are at risk. In addition, as more species arrive, the average population size of each will diminish as competition increases.

At first, the few species present can occupy a greater variety of ecological niches than would be possible on the mainland, where they are competing with many other species. For example, in the comparison mentioned earlier (p. 139) between the Panama mainland and Puercos

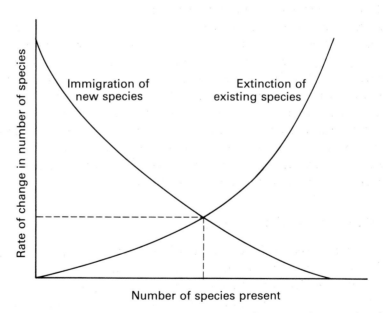

Fig. 6.3 Equilibrium model of the biota of an island (after MacArthur & Wilson [7]). The curve of the rate of immigration of new species and the curve of the rate of extinction of species already on the islands intersect at an equilibrium point. The interrupted line drawn vertically from this point indicates the number of species that will then be present on the island, while that drawn horizontally indicates the rate of change (or turnover rate) of species in the biota when it is at equilibrium.

Island, the smaller number of bird species in the island were able, because of reduced competition, to be far more abundant: there were 1·35 pairs per species per hectare in Puercos Island, compared with only 0·33 and 0·28, respectively, for the two mainland areas [8]. This effect of release from competition was especially noticeable in the antshrike (*Thamnophilus doliatus*). On the mainland, where it competed with over 20 other species of ant-eating bird, there were only eight pairs of antshrike per 40 hectares; on Puercos Island, where there was only one such competitor, there were 112 pairs of antshrike per 40 hectares.

If an island is later colonized by a new species which makes use of foodstuffs similar to those consumed by one of the earlier immigrants, competition between the two species will take place. This may result in the extinction of one of the two competitors, or in the gradual divergence of their food preferences so that the extent to which they are competing with one another is reduced. This latter process, the temporal or spatial separation of species (see pp. 64–67), has the result that each is becoming more specialized in its requirements, making better use of a smaller variety of the possible sources of nourishment. For example, though three different insectivorous species of the tanager (*Tangara*) coexist on the island of Trinidad, competition between them is reduced because they hunt for insects on different parts of the vegetation: *T. guttata* searches mainly on the leaves, *T. gyrola* on the large twigs and *T. mexicana* mainly on the smaller twigs [12]. If the variety of food used by each species is reduced in this way, it must follow that the size of the population of each species that the island can support is now smaller. Since the chances of extinction are greater for smaller populations, the rate of extinction must rise as new species colonize the island, until the equilibrium point is reached at which the rates of colonization and of extinction are equal.

In some situations the rate of extinction may not increase as rapidly as in MacArthur and Wilson's model. For example, if the island had initially contained no fauna or flora at all, there would have been a gradual progression through a number of stages to the climax community. At each stage, some immigrants that would previously have been unable to establish themselves will now, for the first time, find a vacant niche. These changes may also, of course, lead to some extinctions among the earlier colonizing species.

As outlined above, the biotic equilibrium point will also depend upon the size of the island, and on its distance from the source of its colonists. MacArthur and Wilson show this, also, in graphic form (Fig. 6.4). They also give a mathematical analysis of the interaction of these factors, which predicts quite high rates of extinction.

MacArthur and Wilson's Theory of Island Biogeography was widely welcomed, for it gave biogeographers a theoretical background with

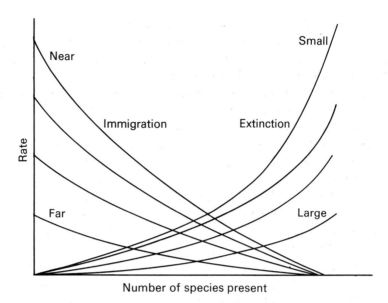

Fig. 6.4 The inter-relationship between isolation and area in determining the equilibrium point of biotic diversity. Increasing distance of the island from its source of colonists lowers the rate of immigration (left). Increasing area lowers the rate of extinction (right). After MacArthur and Wilson [11].

which to compare their own individual results, and therefore encouraged a more structured and less *ad hoc* approach to biogeographical studies. Its methodology has also been extended into types of isolation other than that of dry land surrounded by water. For example, mountain peaks [13], cave biota [14] and individual plants [15] have all been interpreted in this way, and the theory has even been extended to evolutionary time, with individual host-plant species being regarded as islands as far as their 'immigrant' insect fauna is concerned [16,17].

The documentation of extinction, colonization and equilibrium

Though evidence from comparing the biotas of different islands may suggest that they differ in their extinction rates or their immigration rates, and therefore in the potential diversity of their biotas, other studies provide more direct evidence on the validity of these observations. These studies are of the colonization of islands devoid of life, or of islands that have changed in area or have only recently been severed from the mainland.

The most extensive documentation of a natural example of island recolonization is that of the East Indies island of Rakata [18], between the major islands of Java and Sumatra, which act as the sources of its colonists. This island, 17 km² and up to *c.* 780 m high, is the largest remaining fragment of the island of Krakatau, which was destroyed by an enormous volcanic explosion in 1883. All life on Krakatau was extinguished by the eruption, but biotic surveys were made intermittently from 1886, with a gap between 1934 and 1978, and intensive work since the centenary of the eruption, in 1983. These surveys show that the patterns of colonization and extinction are not smooth, but are heavily influenced by the times of emergence of new ecosystems, and by the linkage between plants and animals due to food requirements or mechanisms of dispersal.

Nearly 50 per cent (11/24) of the earliest recorded (1886) species of plant colonists were ferns, whose wind-blown spores arrived readily, and which formed most of the vegetational cover of the interior of the island, though only three species survived the later plant successional changes. Although the plant immigration rate was at first high and then dropped (Fig. 6.5a), it later increased again as the establishment of new ecosystems such as forest provided new available niches. When, in the 1920s, the forest covered the last of the open grassland, there was a new period of extinction of grassland plants (and of grassland butterflies and birds) as their habitat disappeared.

Analysis of the methods of plant dispersal to Rakata throws further interesting light on the history of colonization (Fig. 6.5b). The coastal community, which arrived by sea, was (not surprisingly) the first to become established, and in 25 years (by 1908) formed over 70 per cent of the seed-plants. Few (12 per cent) of the original types of colonizing coastal seed-plants have since become extinct, presumably because their environment has been constant and also frequently supplied with new colonists. The more inland seed-plants have mostly arrived by wind dispersal; 50 per cent of the inland species recorded in 1897 have since become extinct, presumably because of the successional change noted above.

Dispersal by animals became most important from the 1920s, as they were the main source for the larger, heavier seeds of the trees of the inland forests that were spreading and diversifying at that time; 53 of the 57 species of this group are animal-dispersed. There was probably an element of positive feedback in this, as the greater number of fruiting trees attracted more visiting bats and birds, which in turn dropped seeds in their excreta.

Further insight can be derived from the study of the butterflies of Rakata. Since 1933 the number of butterfly species reliant on sea-dispersed plants nearly doubled, although the composition of this plant community has hardly changed over the last 90 years. This suggests that the number of these species is still limited by the problems of access, even in these coastal butterflies, whose habitat makes them the most liable to being blown out to sea and becoming involuntary colonists.

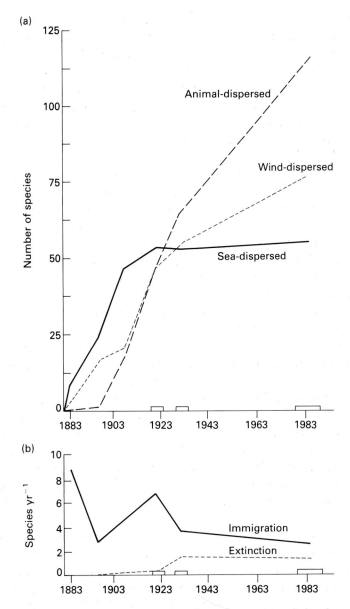

Fig. 6.5 History of the Rakata higher plant flora since 1883, excluding human introductions. (a) modes of dispersal (b) rates of immigration and extinction. After Bush & Whittaker [18].

This in turn suggests that access is also the limiting factor for the butterflies of the more inland regions, whose habitat makes them less liable to being blown out to sea. In the case of the deep forest, however, which has only been in existence for 50–60 years, the problem may instead be lack of food plants rather than merely that of dispersal of the butterflies.

The immigration of new species of forest tree is itself also likely to be slow, as any immigrant must take over ground that is already occupied by an existing species, which is itself long-lived. It will therefore probably take many hundreds of years for this forest to become fully mature, with a range of ages of tree, from saplings to dying trees. Since environmental stability over such a long period of time is unlikely, lowland tropical forests at equilibrium are probably rare or non-existent.

The complexity of all these ecological and successional changes strongly suggests that it is highly unlikely that the colonization history of an island will follow the simple path predicted by the Theory of Island Biogeography. The replacement of one plant community by another will cause pronounced irregularities in the graphs of immigration and extinction, not only of the plants themselves but also of the associated fauna. For example, the bird fauna of Rakata includes the flowerpecker *Dicaeum*, which distributes the seeds of the plant family Loranthaceae, which are epiphytic plant parasites of the trees of the forest canopy. But the Rakata forest is not yet old enough to contain the mature and dying trees that the parasite can attack. As a result the Loranthaceae are absent, together with the butterfly *Delias* that feeds upon these plants, even though the butterfly itself is highly migratory and a competent potential colonist. The integrated nature of the successional ecosystems therefore makes it likely that the colonization history of an island like Rakata will show pronounced waves of change, rather than the simple monotonic curve predicted by the Theory of Island Biogeography.

More detailed studies of the progress of island colonization have been made by Dan Simberloff and Edward Wilson, who fumigated four small islets in Florida Bay to remove the insect population [19]. Their initial studies showed that the rates of insect colonization and extinction of an islet 200 m from its source area were quite high, resulting in a turnover rate of one species every one to two days (the original insect fauna having been between 20 and 50 species, out of a total insect fauna of several hundreds in the Florida Keys area). The insect fauna of the islands had returned to the original level within about six months and, although further colonization and extinction continued thereafter, the number of insect species did not change significantly. This return to the original level of insect diversity (Fig. 6.6) was despite the fact that only 25–30 per cent of the original species were present in the island faunas one year after defaunation. Later, more intensive studies and analyses by Simberloff [20] suggested that the original studies had overestimated the number of extinctions, and that there was really not more than 1·5 extinctions per year. This experiment therefore not only demonstrated quite a high rate of colonization and extinction, but also supported the view that there was an approximate short-term constancy of species numbers and a fair amount of turnover.

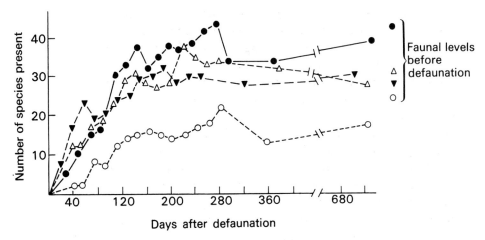

Fig. 6.6 The colonization curves of four small mangrove islets in the Florida Keys. After Simberloff & Wilson [19].

Second thoughts about the Theory

Many papers have now been written that interpret individual biota in terms of the Theory of Island Biogeography. These papers were in turn taken as providing such a wide measure of support for the Theory that it became almost uncritically accepted as a basic truth. In turn, therefore, results which did not conform to expectations based on the Theory were re-examined in search of procedural or logical faults, or for unusual phenomena that might explain this 'anomalous' result, or simply ignored, rather than being seen to cast doubts on the applicability or universality of the Theory. More recently, however, a number of criticisms have been made, both of the nature of the Theory itself and of the extent to which subsequent work can be taken to have 'proved' or supported it. For example, the Theory assumes that there is an equilibrium number of species and that this is affected only by the physical characteristics of island area and degree of isolation. Species are therefore treated as simple numerical units, of equal value to one another. Their possible biological interactions, such as competitive or co-evolutionary effects, are therefore ignored or assumed to be trivial in comparison with the overall statistical effects, as is the possibility of an increase in the number of species by evolution rather than by immigration.

Simberloff's experiments on tiny islands of a single type are virtually the only ones in which detailed, intensive and continuous study over a period of time provide an adequate basis for interpretation. However, Simberloff himself later cast doubt on the significance of these results, stating that the data demonstrated mainly a 'pseudoturnover' caused by the movement of transient insects from island to island, and not a turnover of a resident population due to an immigration/extinction equilibrium [20]. As we have already seen in the case of Rakata, an island biota is not merely an assemblage of individual communities but is also, over a period of time, a dynamic system in which new communities arise out of older ones. Similarly, Thomas Schoener and David Spiller, who studied the populations of eight species of orb spider in 108 Bahamas islands over an eight-year period, found that larger, permanent populations appear to be the only source of colonizing individuals for a subset of smaller, fugitive populations that are relatively unimportant in biomass or longer-term ecological significance [21]. The total island biota is therefore again seen as a complex of interconnected units.

Many studies which have been extensively quoted as supporting the Theory of Island Biogeography are really far too imprecise, as pointed out by Simberloff in his later paper [20] and in a review by the British biologist Francis Gilbert [22], while the statistical procedures of many earlier studies were criticized by the American ecologists Edward Connor and Earl McCoy [23]. It should also be realized that the Theory's prediction that the number of species will remain at an equilibrium despite both immigration and extinction refers to a situation in which the environment remains relatively constant. It would be expected that major ecological changes would alter the fundamental balance, so that 'before' and 'after' comparisons would have little relevance to the predictions of the Theory in which the environment remains relatively constant (see also p. 161−2).

For example, Jared Diamond [24] compared the numbers of bird species found in the Californian Channel Islands in a 1968 survey with those recorded in a 1917 review. He concluded that there had been an equilibrium in the number of species breeding there, even though the turnover had been as high as 20−60 per cent of the species on each island. (Surprisingly, this turnover was related to faunal size, being greater on islands with fewer species and, contrary to the predictions of the Theory, was independent of either the size of the island or its degree of isolation.) However, it was later pointed out by Lynch and Johnson [25] that Diamond had ignored the fundamental changes in the environment of the Channel Islands that had taken place between 1917 and 1968. These included the effects of human activities, such as hunting and the poisoning of birds of prey by pesticides, and the appearance of totally new species, such as sparrows, in California. The '1917' baseline was, in any case, a summation of records covering some 50 years. Lynch and Johnson also found similar flaws in other studies of bird faunas on an island near New Guinea carried out by Diamond [26], and on a West Indies island by Terborgh and Faaborg [27]. The conclusions based on analysis of other bird faunas by Jones and Diamond [28] and Diamond and May [29], which show considerable turnover rates, are weakened by the fact that these changes particularly involve species for whom the very short sea crossings present no real barrier, so that for them the areas of land are islands only in a theoretical sense. Finally, Simberloff [30] analysed the records of the bird faunas of two islands and three inland areas over 26–33 years, and found that none of them showed any evidence of regulation towards an equilibrium.

Over even greater periods of time, the quality and adequacy of the data become even more unreliable. For example, the American botanist Beryl Simpson [31] studied the flora of the high-altitude patches of 'paramo' vegetation on Andean mountains. She concluded that their diversity could be correlated better with the deduced, smaller areas that they would have occupied during the Pleistocene lower temperatures than with their areas today. Similarly, Jared Diamond [32] suggested that the bird faunas of some Pacific islands could best be interpreted as in the process of reducing (or 'relaxing') towards a new, lower balance because rising sea levels had reduced the areas of the islands. The problem here is that the estimated required times for these biotic changes are so long that, almost inevitably, further geographical or climatic events will have changed the situation afresh. If so, this casts doubt on the extent to which any of these island biotas can be thought of as being 'at equilibrium'. This in turn weakens the case for viewing any of the data as providing a sufficiently reliable base for the Theory of Island Biogeography.

The theory is probably on firm ground when it attempts to provide a unifying theory that describes the interactions of a large number of largely independent variables, such as island area, isolation, habitat diversity and position. The problem arises when it also suggests that their interactions are sufficiently comprehensible and regular that a simple model can result, with which observations of new situations can be compared and which can illuminate those results. Dan Simberloff, who has become one of the most analytical critics of the Theory, believes that its application in any one example has to be so modified by individual circumstances that it cannot be disproved by any observation or statistical analysis [33]. The Theory has ceased to be a general abstraction of the elements common to all island communities, and become merely a framework of narrative descriptions of each situation.

It has also been pointed out that the patterns of interaction predicted by the Theory should be tested against an alternative 'null model' that generates a different set of patterns. However, as Robert Colwell and David Winkler have shown [34], it is in fact extremely difficult to design a null model that does not itself suffer from serious biases. It is equally difficult to use a real-world mainland biota as a comparison, because it is itself the result of competition, both in the past and in the present, that has conditioned its composition.

The Theory of Island Biogeography and the design of nature reserves

It is not surprising that the Theory of Island Biogeography was warmly welcomed by those concerned with the management or design of nature reserves, for they could be analysed as islands amidst the surrounding unprotected land. It seemed to promise an almost magical prescription for ensuring the ideal balance between effectiveness in terms of retaining the maximum number of species, and economy in their area (and therefore in their cost of acquisition and management). After the realization of the limitations of the Theory, as reviewed above, it is now clear that it is no such remedy and reliable forecaster. As Shafer [35] has commented, no aspect of the concept of an equilibrium level in species numbers can now be considered as established, and this would in any case be of limited value in an environment that is subject to both cyclical and occasional change.

Only one of the predictions of the Theory, that the number of species in an island will increase directly with enlargement in its area, now seems reliable. Large nature reserves should therefore retain more species and suffer fewer extinctions. But how large need it be? Very little is known about the area requirements of most species, and these will depend on their degree of ecological specialization and on their breeding habits. The larger mammals and birds that are a major preoccupation of many tropical game reserves require a very large home range to meet their energy needs. But, within the size range of the world's nature reserves, 97·9 per cent of which are less than $10\,000\,km^2$ in area, the variability of prediction of the Theory is so great as to be of no practical value [36]. Shafer comments that it is probably impossible to separate the effects of the area of a nature reserve from other factors, even using sophisticated statistical techniques of multivariate analysis. On the other hand, the Theory of Island Biogeography, together with all the research and analysis that the Theory has provoked, does underline the fact that it is not only the size of the nature reserve itself that is important, but also the preservation of the individual habitats within it at as large a size as possible, and the preservation of links with other similar areas, so as to maximize the chances of inward colonization.

Opportunities for adaptive radiation

Colonists may encounter many difficulties when they first enter an island, but there are rich opportunities for those species that can survive them long enough for evolution to adapt them to the new environment. These opportunities exist because of the lack of many of the parasites and predators that elsewhere would prey upon the species, and of many of the other species with which it normally competes. Like Darwin's finches on the Galápagos Islands, it may be able to radiate into ways of life not formerly available to it.

A good example of this can be found in the Dry Tortugas, the islands off the extreme end of the Florida Keys, which only a few species of ant have successfully colonized [11]. One species, *Paratrechina longicornis*, on the mainland normally nests only in open environments under, or in the shelter of, large objects; but on the Dry Tortugas it also nests in environments such as tree trunks and open soil, which on the mainland are occupied by other species. Not every species, however, is capable of taking advantage of such opportunities in this way. On the Florida mainland, the ant *Pseudomyrmex elongatus* is confined to nesting in red mangrove trees, occupying thin hollow twigs near the tree top. Though it has managed to colonize the Dry Tortugas, it is still confined there to this very limited nesting habitat.

Such opportunities for alterations in behavioural habits or in diet provide in turn the opportunity for the organism to become permanently adapted, by evolutionary change, to a new way of life. This process requires a longer period of time, and is therefore unlikely to take place except on islands that are large and stable enough to ensure that the evolving species does not become extinct. But if an island does provide these conditions, then remarkable evolutionary changes may take place as colonizing species become modified to fill vacant niches. Instances of such islands are provided also by the isolated Great Lakes of Africa, large enough to provide a great diversity of environments, and old enough for these ecological opportunities to have become realized through evolutionary change. The cichlid fishes, in particular, have been able to take advantage of this, and have undergone a rapid evolutionary change, producing 37 genera and 126 species in Lake Tanganyika, and 20 genera and 196 species in Lake Malawi [37]. Most other lakes are probably too small, too impermanent, or too recent in origin, for colonization or evolutionary change to have been able to replenish the losses due to extinction, and are faunally impoverished. In particular, it probably takes a considerable time for the deeper parts of lakes to become exploited by evolutionary change of the normally shallow-water fishes that are the only possible colonists.

One of the opportunities that may exist on an island often results from absence of one of the normal elements of a mainland flora — the tree. The seeds of trees are usually much larger and heavier than those of other plants and are therefore not readily transported long distances. As a result, other plants may develop to fill this vacant niche [38]. The modifications needed to produce a tree from a shrub which already possesses strong, woody stems are comparatively slight — merely a change from the many-stemmed, branching habit to concentration on a single, taller trunk. For example, though most members of the Rubiaceae are shrubs, this family has produced on Samoa the 8 m tall tree *Sarcopygme*, which has a terminal palm-like crown of large leaves. Though more comprehensive changes are needed to produce a tree from a herb, many islands show examples of this phenomenon. In many cases the plants involved are members of the Compositae, perhaps because they have unusually great powers of seed dispersal, are hardy and often already have partly woody stems. To this family belong both the lettuces, which have evolved into shrubs on many islands, and the sunflowers. On the isolated island of St Helena in the South Atlantic can be found five different trees, 4–6 m high, which have evolved in the island from four different types of immigrant sunflower (Fig. 6.7). Two of these (*Psiadia* and *Senecio*) are endemic species of more widely distributed genera, while the other three (*Commidendron*, *Melanodendron* and *Petrobium*) are recognized as completely new genera. These latter genera are therefore both endemic — that is, known only in St Helena — and also *autochthonous* — that is, they actually evolved in the area concerned.

A similar process is responsible for the appearance of large animals on some islands: for example the Komodo dragon (*Varanus komodoensis*), a giant lizard which lives on Komodo Island and nearby Flores Island in the East Indies. These animals increased in size to occupy niches which on the mainland are filled by animals much larger than typical varanid lizards.

All these organisms evolved in islands to fill habitats normally closed to them. But other evolutionary changes frequently found on islands are the direct result of the island environment itself, not of the restricted fauna and flora. We have seen how serious may be the effect of a small population. But the same island will be able to support a larger population of the same animal if the size of each individual is reduced. This evolutionary tendency on islands is shown by the find of fossil pygmy elephants that once lived on islands in both the Mediterranean and the East Indies. On a much smaller scale, the size of lizards on four of the Canary Islands still shows the same phenomenon. The head to vent length of males of the lizard *Lacerta galloti* ranges from 135 mm on the largest island, Tenerife, to only 82 mm on Hierro, the smallest.

Another tendency is for island species to lose the very dispersal mechanisms that allowed them to reach their home. Once on the restricted area of the island, the ability for long-distance dispersal is no longer of value to the species: in fact it is a disadvantage. The seeds of plants tend to lose their 'wings' or feathery tufts, and many island insects are wingless. The loss of wings by some island birds may be partly for this reason, and partly because there are often no predators from which to escape. A few out of many examples are the kiwi and moa of New Zealand, the elephant birds of Madagascar, and the dodo of

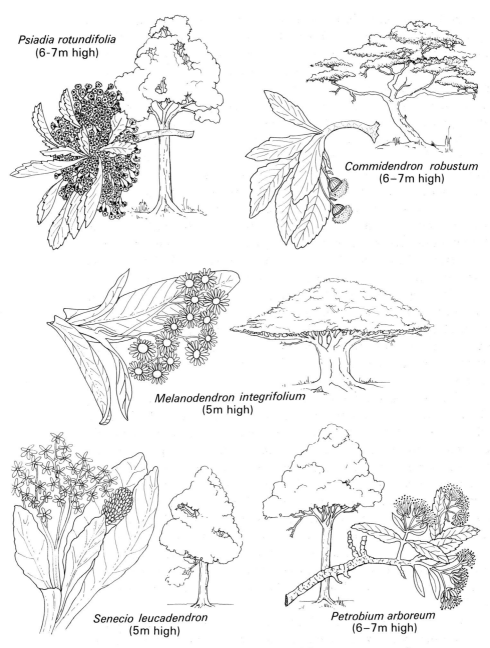

Psiadia rotundifolia
(6-7m high)

Commidendron robustum
(6–7m high)

Melanodendron integrifolium
(5m high)

Senecio leucadendron
(5m high)

Petrobium arboreum
(6–7m high)

Fig. 6.7 The varied trees that have evolved from immigrant sunflowers on St Helena Island. From Carlquist [38].

Mauritius (the last three are extinct, but only because man was the predator).

The Hawaiian Islands

As has been seen, there are many aspects of island life that are unique, and many others that differ only in degree from life on the continental land masses. The result of the action of all these different factors can be seen by examining the flora and fauna of one particular group of islands. The Hawaiian Islands provide an excellent example, for they form an isolated chain, 2650 km long, lying in the middle of the North Pacific, just inside the Tropics (Fig. 6.8). Sherwin Carlquist has provided an interesting account of the islands and of their fauna and flora, pointing out the significance of many of the adaptations found there [39].

The islands are of volcanic origin, rising steeply from a sea floor which is 5500 m deep, to the volcanic peaks which reach up to 4250 m. Hawaii itself lies 3200 km from North America and 5500 km from Japan. The islands seem to be the result of the activity of a particular point in the earth's interior, past which the sea floor has been moving westwards as the Pacific plate moves in that direction (see Chapter 7 and Fig. 7.1).

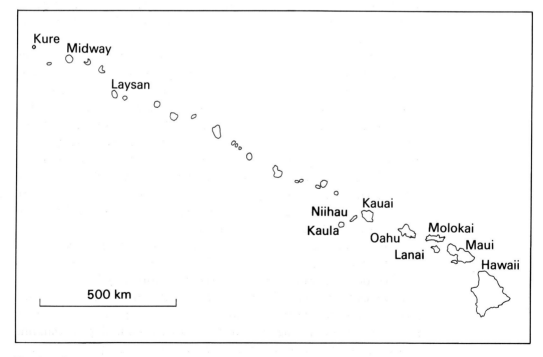

Fig. 6.8 The Hawaiian Island chain.

The most easterly, Hawaii, is therefore the youngest (less than 700 000 years old) and bears the still active volcanoes Kilauea and Mauna Loa. The most westerly island still visible, Kure, is about 15 million years old, but the Emperor Seamount chain, consisting of the remains of other islands now eroded to below sea level, extends further westwards and then northwards to near the Kamchatka peninsula of Siberia. The oldest of these submerged seamounts is over 70 million years old, so that there must have been a group of islands in the north-central Pacific for at least this length of time.

It is clear that animals and plants have been able to colonize islands in this region for a very long period of time. Once arrived, they have been able to diversify and, often, to spread to the newly formed additions to the chain as the older islands steadily eroded away and disappeared beneath the sea. The high volcanic peaks seem to be islands within islands, for it is difficult for their alpine plants to disperse from one island to another. Instead, they have evolved from the adjacent lower-lying flora at each island: 91 per cent of the alpine plants of Hawaii are endemic, a far higher proportion than that of the island flora as a whole (16·5 per cent) or even of the angiosperms alone (20 per cent) [40].

The most obvious result of the extreme isolation of the Hawaiian Islands is that many groups of animal are completely absent. There are no truly freshwater fish and no native amphibians, reptiles or mammals (except for one species of bat). The difficulty of getting to the islands is shown by the fact that the present-day bird fauna of Hawaii seems to be the result of only 15 different colonizations.

The closest relatives of most of the Hawaiian animals and plants live in the Indo-Malaysian region. For example, of the 1729 species and varieties of Hawaiian seed plants, 40 per cent are of Indo-Malaysian origin but only 18 per cent are of American origin; also, nearly half of the 168 species of Hawaiian ferns have Indo-Malaysian relatives, but only 12 per cent have American affinities. This is not surprising, for the area to the east is almost completely empty, while that to the south and west of the Hawaiian chain contains many islands, which can act as intermediary homes for migrants. Organisms adapted to life in these islands would also be better adapted to life in the Hawaiian Islands than would those from the American mainland.

The way in which the Hawaiian birds reached the islands is obvious enough. One of the plants which probably came with them is *Bidens*, a member of the Compositae, whose seeds are barbed and readily attach themselves to feathers; about 7 per cent of the Hawaiian non-endemic seed plants probably arrived in this way [5]. The Hawaiian insects, too, arrived by air. Entomologists have used aeroplanes and ships to trail fine nets over the Pacific at different heights and have trapped a variety of insects, most of which, as would be expected, were species with light bodies. These types also predominate in the Hawaiian Islands (an indication of their air-borne arrival) though heavier dragonflies, sphinx-moths and butterflies are also found there.

The influence of the winds in providing colonists is shown by the fact that, though angiosperms are far more common than ferns in the world as a whole, their diversity in Hawaii is more evenly balanced — 225 immigrant angiosperms, 135 immigrant ferns. The relatively greater success of the ferns is probably due to the fact that their spores are much smaller and lighter than the seeds of angiosperms. Of the non-endemic seed plants of the Hawaiian Islands, about 7·5 per cent almost certainly arrived carried by the wind, while another 30·5 per cent have small seeds (up to 3 mm in diameter) and may also have arrived in this way.

One of the most interesting plants that probably arrived as a wind-borne seed is the tree *Metrosideros*. It is unusual because its seeds are tiny compared with those of other trees, and this has allowed it to become widely dispersed through the Pacific Islands. It is a pioneering tree, able to form forests on virtually soilless lowland lava rubble — a great advantage on a volcanic island. *Metrosideros* shows great variability in its appearance in different environments, from a large tree in the wet rain forest, a shrub on wind-swept ridges, to as little as six inches high in bogs, and it is therefore the dominant tree of the Hawaiian forest. Though these differences are probably at least partially genetically based, the different forms are not distinct species, and intermediates are found where two different types (and habitats) are adjacent to one another.

Probably the single most important method of entry of seed plants to the Hawaiian Islands has been as seeds within the digestive system of birds that have eaten fruit containing the seed (e.g. blueberry, sandalwood); about 37 per cent of the non-endemic seed plants of the islands probably arrived in this way. Significantly, many plants that succeeded in reaching the islands are those which, unlike the rest of their families, bear fleshy fruits instead of dry seeds (e.g. the species of mint, lily and nightshade found in Hawaii).

Dispersal by sea accounts for only about 5 per cent of the non-endemic Hawaiian seed plants. As well as the ubiquitous coconut, the Islands also contain *Scaevola taccata*; this shrub has white buoyant fruits and forms dense hedges along the edge of the beach on Kauai Island. Another sea-borne migrant is *Erythrina*; most species of this plant genus have buoyant bean-like seeds. On Hawaii, after its arrival on the beach, *Erythrina* was unusual in adapting to an island environment, and a new endemic species, the coral tree *E. sandwichensis*, has evolved on the island. Unlike those of its ancestors, the seeds of the coral tree do not float — an example of the loss of its dispersal mechanism often characteristic of an island species.

The successful colonists of the Hawaiian Islands are the exceptions; many groups have failed to reach them. As already noted, no terrestrial vertebrates occur there naturally, while 21 orders of insect are completely absent. As might be expected, most of these are types that seem in general to have very limited powers of dispersal. For example, the Formicidae (ants), which are an important part of the insect fauna in other tropical parts of the world, were originally absent for this reason. They have, however, since been introduced by man, and 36 different species have now established themselves and filled their usual dominant role in the insect fauna. This proves that the obstacle was entrance to the islands, not the nature of the Hawaiian environment.

As ever, the absence of some groups has provided greater opportunities for the successful colonists. Several insect families, such as the crickets, fruit-flies and carabid beetles, are represented by an extremely diverse adaptive radiation of species, each radiation derived from only a few original immigrant stocks [41]. The fruit-flies, belonging to the closely related genera *Drosophila* and *Scaptomyza*, have been studied in particular detail [42]. Of all the known species (over 1200) about one-third are found only in the Hawaiian chain. The fact that the island of Hawaii is less than 700 000 years old also implies that the species on that island must have differentiated within that time. Detailed studies of the chromosome structure of the Hawaiian fruit-flies are now even making it possible to reconstruct the sequence of colonizations which must have taken place [43]. As might have been expected, the older islands to the west in general contain species ancestral to those in the younger islands in the east (Fig. 6.9). The youngest, Hawaii itself, has nineteen species descended from species in older, more westerly islands, but none of its species appear to be ancestral to those in the older islands. (The present-day islands of Maui, Molokai and Lanai together formed a single island until the post-glacial rise in sea-levels; their *Drosophila* species are therefore treated together as a single fauna.)

The abundance of species of fruit-flies in these islands is probably due partly to the great variations in climate and vegetation to be found there, and also the periodic isolation of small islands of vegetation by lava flows. But another major factor has been that the Hawaiian fruit-flies, in the absence of the normal inhabitants of the niche, have been able to use the decaying parts of native plants as a site in which their larvae feed and grow. This change is probably also due to the fact that their normal food of yeast-rich fermenting materials is rare in the Hawaiian Islands.

The same phenomenon of a great adaptive radiation has taken place in other groups of animals and plants. In general, therefore, though the islands contain comparatively few different families, each contains an unusual variety of species, nearly all of which are unique to the islands.

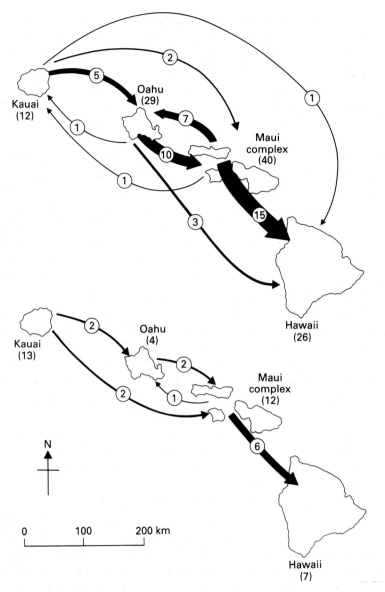

Fig. 6.9 The dispersal events within the Hawaiian Islands that are suggested by the inter-relationships of the species of 'picture-winged' *Drosophila* flies (above) and of tarweed plant (below). The width of the arrows is proportional to the number of dispersal events implied, and the number of species in each island is shown in parentheses. From Carr *et al.* [44], by permission of Oxford University Press.

In fact, out of the whole of the Hawaiian flora, over 90 per cent of the species are endemic to the islands.

There are many other examples of Hawaiian adaptive radiations, but three are of particular interest: the tarweeds and lobeliads among the plants [39], and the honey-creepers among the birds. The tarweeds belong to the family Compositae, and probably arrived on the islands as sticky seeds attached to the feathers of birds. They have produced only three genera in the islands (*Dubautia*, *Argyroxiphium* and *Wilkesia*), but these have colonized a variety of habitats. For example, on the bare cinders and lava of the 3050 m high peak of Mt Haleakala on the island of Maui, two of the few plant species that can survive are tarweeds. *Dubautia menziesii* is adapted to this arid environment by its tall stem and stubby succulent leaves, while the silversword *Argyroxiphium sandwichense* is covered by silvery hairs that reflect the heat. A few hundred metres below the bare volcanic peaks, conditions are at the other extreme, because most of the rain falls at heights of from 900 to 1800 m; these regions receive from 250 to 750 cm of rain per year. The upper regions of 1770 m high Mt Puu Kukiu on Maui are covered by bog in which thrives another silversword, *A. caliginii*. On the island of Kauai the heavy rainfall has led to the development of dense rain forest, in which *Dubautia* has evolved a tree-like species, *D. knudsenii*, with a trunk 0·3 m thick, and large leaves to gather the maximum of sunlight in the dim forest. Kauai bears another tarweed which shows the tendency for island plants to become trees. In the drier parts of this island grows *Wilkesia gymnoxiphium*, with a long stem which carries it above the shrubs that compete with it for light and living space. This species also shows another example of the loss of the dispersal mechanism that first brought the ancestral stock to the island: the seeds of *Wilkesia* are heavy and lack the fluffy parachutes usually found among the Compositae. The pattern of dispersal of the 28 species in the three tarweed genera within the Hawaiian Islands is very similar to that of the drosophilid flies (Fig. 6.9) [44].

Lobeliads (members of the plant family Lobeliaceae) are found in all parts of the world, but they have undergone an unusual adaptive radiation in the Hawaiian Islands, because their normal competitors, the orchids, are rare. The Hawaiian lobeliads include 150 endemic species and varieties, making up six endemic genera. Over 60 species of one endemic genus alone (*Cyanea*) are known, showing an incredible diversity of leaf form (Fig. 6.10). The plants range from the tree *C. leptostegia*, 9 m tall (similar in appearance to the tarweed *Wilkesia*) to soft-stemmed *C. atra*, only 0·9 m tall. The species of another genus, *Clermontia*, are less varied in overall size, but are very varied in the size, shape and colour of their flowers. These are mainly tubular and brightly coloured, a type of flower that is often associated with pollination by birds. On isolated islands such as Hawaii the adaptation of larger flowers to pollination by birds may be because the large insects that would normally pollinate such flowers on the mainland are absent. It is no coincidence that the adaptive radiation of the Hawaiian lobeliads has been accompanied

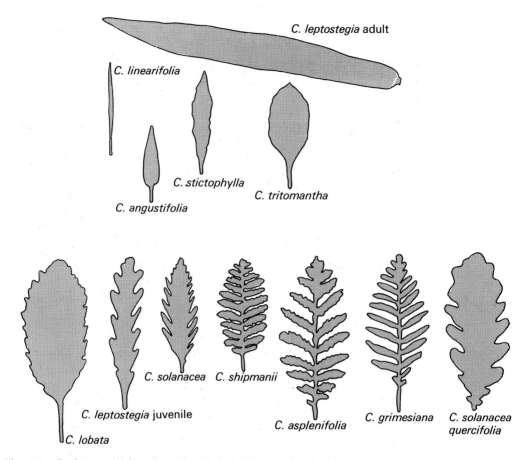

Fig. 6.10 The leaves of different species of the lobeliad genus of *Cyanea*.

by the adaptive radiation of a nectar-eating type of bird, the honey-creepers [45].

The ancestors of these birds was probably a finch-like immigrant from Asia [46] which fed on insects and nectar. From the original immigrants, adaptive radiation has produced 11 endemic genera comprising the endemic family Drepanididae (Fig. 6.11). Many of the genera, such as *Himatione*, *Vestiaria*, *Palmeria*, *Drepanidis*, many species of *Loxops*, and one species of *Hemignathus* (*H. procerus*) are still nectar-eaters, feeding from the flowers of the tree *Metrosideros* and the lobeliad *Clermontia*. Since insects, too, are attracted to the nectar, it is not surprising to find that many nectar-eating birds are also insect-eaters, and from this it is a short step to a diet of nothing but insects. *Hemignathus wilsoni* uses its mandible, which is slightly shorter than the upper half of its bill, to probe into crevices in bark for insects, and *Pseudonestor xanthophrys* uses its heavier bill to rip open twigs and branches in search of insects. Species of the genus *Psittacirostra* too,

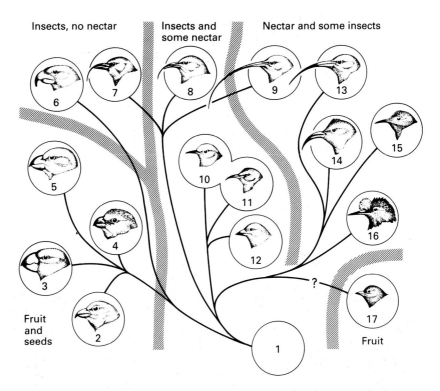

Fig. 6.11 The evolution of dietary adaptations in the beaks of Hawaiian honey-creepers. 1, Unknown finch-like colonist from Asia. 2–5, *Psittacirostra psittacea*, *P. kona*, *P. bailleui*, *P. cantans*. 6, *Pseudonestor xanthophrys*. 7–9, *Hemignathus wilsoni*, *H. lucidus*, *H. procerus*. 10–12, *Loxops parva*, *L. virens*, *L. coccinea*, 13, *Drepanidis pacifica*. 14, *Vestiaria coccinea*. 15, *Himatione sanguinea*. 16, *Palmeria dolei*. 17, *Ciridops anna*.

have heavy, powerful beaks, which some use for cracking open seeds, nuts or beans. The light bill of the recently extinct *Ciridops* was used for eating the soft flesh of the fruits of the Hawaiian palm *Pritchardia*.

Recent studies of the Hawaiian avifauna also show the unreliability of modern biota as a basis for estimates of rates of biotic change or of the relationship between island area and the number of species. It has long been known that about a dozen Hawaiian bird species became extinct after the arrival of Europeans and the animals that they introduced. But studies by the American biologists Storrs Olson and Helen James [47,48] revealed at least 50 now-extinct species of Hawaiian bird — more than the entire avifauna of the Islands today. These included flightless types of ibis, rail and goose-like ducks, six hawks and goshawk-like owls, and nearly two dozen species of drepanidid finches, mostly of insectivorous type. Most of these species were still alive when the Polynesians arrived in the Islands in about AD 300, but had become extinct before Europeans arrived 1500 years later. Similar evidence for bird extinctions has been reported from many other Pacific islands.

Living as they do on fruits, seeds, nectar and insects, it is not surprising that none of the drepanidids shows the island-fauna characteristic of loss of flight. However, both on Hawaii and on Laysan to the west, some genera of the Rallidae (rails) have become flightless (a common occurrence in this particular family of birds). The phenomenon of flightlessness is also common in Hawaiian insects: of the endemic species of carabid beetle, 184 are flightless and only 20 are fully winged. The Neuroptera or lacewings are another example — their wings, usually large and translucent, are reduced in size in some species, while in other species they have become thickened and spiny.

To summarize, islands provide a unique opportunity to study evolution, for their small, impoverished faunas and floras are the ideal situation for rapid evolutionary modification and adaptive radiation. At the same time, island life is unusually hazardous, so that there is a complex interaction between the processes of immigration, colonization and extinction; quantitative analysis of this has recently commenced, and is proving extremely valuable in providing insights into the history and structure of ecosystems, on the continents as well as on islands.

Drawing lines in the oceans

Though, at first sight, the conditions of life on islands seem totally different from those in the oceans that surround them, in fact the faunas of the shelf regions that border the oceans are almost as isolated. In general, the faunas of these shelf regions, which descend to depths of about 200 m, follow the broadly latitudinal bands of temperature change, becoming more diverse as they approach the equator. The American biologist John Briggs [49] has attempted to identify the zones where the rates of faunal change are most rapid (Fig. 6.12), and he notes that the northern and southern boundaries of the tropical faunas appear to be most closely related to the 20°C isotherm for the coldest month of the year. In the warm-temperate regions the average temperatures for the coldest month are normally 13–20°C (though in the North Atlantic the lowest average temperature may be only 10°C), in the cold-temperate regions 2–13°C, and in the cold Arctic–Antarctic regions from −20°C to +20°C.

Although broadly latitudinal, the boundaries between these marine faunal regions are also affected by the surface currents, which in turn are controlled by wind directions. So, because the major wind circulation forms anticlockwise cells in the Southern Hemisphere and clockwise cells in the Northern Hemisphere (see p. 80), the boundaries between the marine faunal regions are shifted in corresponding directions. This

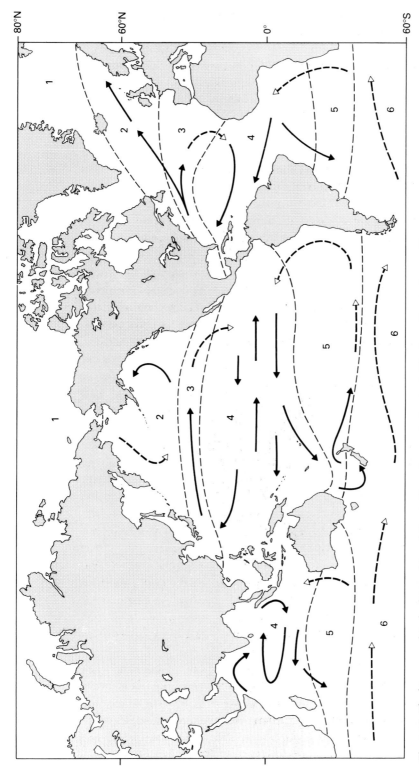

Fig. 6.12 The biogeographical relationships of the shelf-sea faunas. 1, Arctic; 2, Northern Hemisphere cold-temperate; 3, Northern Hemisphere warm-temperate; 4, Tropical; 5, Southern Hemisphere warm-temperate; 6, Southern Hemisphere cold-temperate and Antarctic. Open-headed arrows indicate cold ocean surface currents. Solid-headed arrows indicate warm ocean surface currents. After Briggs [49].

is most marked in the eastern Pacific where, because South America extends far south into cold-temperate Antarctic waters, a cold 'Peru Current' runs far northwards up the western coast of that continent.

It is also interesting to note that, compared with the cold-temperate marine fauna of the North Pacific, that of the North Atlantic is much less diverse and has fewer endemic forms. This may be because, during the Ice Ages (see Chapter 10), ice extended much further south in the Atlantic than in the Pacific, perhaps because the Atlantic is more widely connected to the cold waters of the Arctic Ocean, and is a smaller ocean within which temperature changes will be less buffered by contact with a large body of warmer, more southerly, water.

However, it is important to realize that these marine regions are not like those established for terrestrial animals and plants (see Figs 8.1, 8.3), each of which is defined by having many similar or endemic taxa that occur widely over most of the region. For terrestrial regions, that has been made possible by the fact that these land masses have, to a greater or lesser extent, had independent histories as isolated continents within each of which a unique interplay of evolutionary and climatic events took place. No such long-isolated body of ocean has ever existed. Had the great north–south land mass of the Americas been an ancient, continuous barrier between the Pacific and the Atlantic, the tropical faunas of these two oceans might have been able to diversify independently and become distinctively different. However, the Panama land bridge was only completed a few million years ago, and the marine faunas on either side of the Panama isthmus are still very similar to one another. Study of the fossil plankton faunas from either side of the Isthmus provide detailed evidence of the progress of formation of the link, which led to restriction in the circulation of the intermediate depth of water 6·2 million years ago, and of the surface water two million years later [50]. The marine faunal provinces on either side started to become separate 2·4 million years ago, when the land link became complete; this coincided with a drop in sea levels.

The shelf areas that are linked together as, for example, parts of the Northern Hemisphere warm-temperate region, are therefore basically merely an assemblage of shelf areas that share a particular temperature regime, because they are connected by an ocean surface current of that temperature range. This current may carry marine organisms in its waters as it crosses the deep ocean that separates the two shelves on either side of that ocean, but that does not seem to take place as frequently as one might expect. On the contrary, the deep ocean often acts as a very effective barrier. For example, the wide, almost island-free expanse of the East Pacific has long been recognized as a barrier to the dispersal of shelf organisms across the Pacific Ocean. Of the shore fishes that are found either in the Hawaiian Islands or between Mexico and

Peru at the eastern end of the Northern Hemisphere warm-temperate region, only 6 per cent are found in both. Similarly, the Swedish biologist Sven Ekman [51], who recognized and named the East Pacific Barrier, showed that only 2 per cent of the 240 species and 14 per cent of the 11 or 12 genera of echinoderms found in the Indo-West Pacific area had been successful in reaching the west coast of the Americas (the greater success of the genera being because they have been in existence for longer than the individual species, and therefore had a greater length of time in which to cross the Barrier).

The great antiquity of the East Pacific Barrier has been shown by Richard Grigg and Richard Hey of the University of Hawaii [52], who have recently studied the zoogeographic affinities of fossil and living genera of coral. They found that those of the East Pacific are more closely related to those of the West Atlantic than to those of the West Pacific, even for corals living as long ago as the Cretaceous Period (see Fig. 7.2), proving that the Barrier was effective in inhibiting dispersal across the Pacific as long ago as that time. Shelf faunas also provide evidence on the progressive widening of the Atlantic Ocean. Calculation of the degree of similarity of the shelf faunas on either side of the North Atlantic from the Early Jurassic onwards, using the coefficient of faunal similarity (see p. 184) shows a steady decrease in their similarity as they are gradually separated by the widening ocean [53].

Some genera of the shelf fauna are found in both the Northern and the Southern Hemisphere temperate zones but not in the intervening warmer waters; this is known as a 'bipolar' or 'antitropical' distribution. Various explanations have been put forward for this phenomenon. Darwin suggested that the cooling of the equatorial waters during the Ice Ages had allowed these genera to pass through waters that are now once again too warm for them to inhabit. More recently, Brian White [54] has argued that it was instead the warmer temperatures of the Miocene Epoch that made it impossible for these genera to live in the equatorial waters, so that they now show a relict distribution on either side of this zone. Yet another explanation is that of John Briggs [55], who suggests that these genera have become extinct in the equatorial regions because of competition from newly evolved genera, originating in the rich, faunally diverse Indo-West Pacific region of the tropics. Evaluation of these two contrasting explanations will demand a better palaeontological record of the history of the antitropical groups than is at present available — and, of course, each explanation may be true in some cases but not in others.

The shelf faunas live in one zone of the system of zones that marine biologists recognize, and that are delimited by the changing physical conditions as one moves downwards in the oceans, away from the light and warmth of the surface. The faunas that live upon the bottom, like

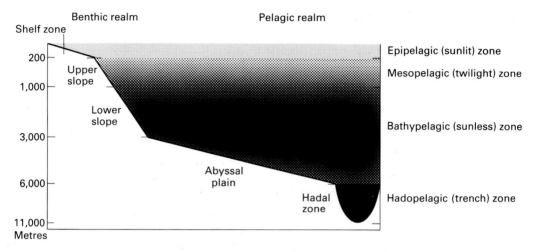

Fig. 6.13 Diagram of the vertical divisions of the ocean, based on distribution of the fauna. After Briggs [49], and reproduced with the permission of McGraw-Hill Inc.

the shelf faunas, are known as 'benthic' faunas, while those that exist within the open ocean are known as 'pelagic' faunas (Fig. 6.13).

The shelves are those parts of the continents that are covered by the seas, and lie at depths of up to 200 m below the surface. At the edges of these continental shelves a steep slope descends to the level of the ocean floor. The 'upper slope' benthic fauna, from 200 to 1000 m depth, occupies the region within which the temperature decreases fairly rapidly with increasing depth. Below this, from 1000 to 3000 m depth, lives the poorer, sparser 'lower slope' fauna. The ocean floor itself, from 3000 to 6000 m below the surface, is the environment of the 'abyssal' fauna, while the sparse fauna of the deep ocean trenches, where old ocean floor disappears back into the depths of the earth (see p. 172) is known as the 'hadal' fauna.

Out in the open ocean, the upper 200 m is the surface layer of the ocean, within which the sunlight can penetrate and in which the temperature is similar to that at the surface; here a rich 'epipelagic' fauna exists. The 'mesopelagic' fauna lives in the twilight zone between 200 and 1000 m, where the temperature rapidly decreases with depth. The fish that live in this zone move upwards and downwards daily according to the time of day. At greater depths, from 1000 to 3000 m, the temperature decreases only slowly, from 5°C to 1.5°C. The fish of the deep 'bathypelagic' zone, from 1000 to 6000 m depth, are too far from the surface for it to be practical to migrate towards the surface. The 'hadopelagic' faunas of the deep trenches are mainly endemic, having evolved within these isolated environments.

References

1 Ridley, H.N. (1930) *The Dispersal of Plants throughout the World*. Reeve & Co., Ashford.

2 Porter, D.M. (1976) Geography and dispersal of Galápagos Islands vascular plants. *Nature* **264**,745–746.

3 Whitehead, D.R. & Jones, C.E. (1969) Small islands and the equilibrium theory of insular biogeography. *Evolution* **23**,171–179.

4 Johnson, M.P., Mason, L.G. & Raven, P.H. (1968) Ecological parameters and plant species diversity. *American Naturalist* **102**,297–306.

5 Van Balgooy, M.M.J. (1971) Plant-geography of the Pacific as based on a census of phanerogam genera. *Blumea*, Supplement **6**,1–222.

6 Mayr, E. (1933) Die Vogelwelt Polynesians. *Mitteilungen Zoologischer Museums Berlin* **19**,306–323.

7 MacArthur, R.H. & Wilson, E.O. (1963) An equilibrium theory of island biogeography. *Evolution* **17**,373–387.

8 MacArthur, R.H., Diamond, J.M. & Karr, J. (1972) Density compensation in island faunas. *Ecology* **53**,330–342.

9 Lack, D. (1969) Subspecies and sympatry in Darwin's finches. *Evolution* **23**, 252–263.

10 Simberloff, D.S. (1978) Using island biogeographic distributions to determine if colonization is stochastic. *American Naturalist* **112**,713–726.

11 MacArthur, R.H. & Wilson, E.O. (1967) *The Theory of Island Biogeography*. Princeton University Press, Princeton.

12 Snow, B.K. & Snow, D.W. (1971) The feeding ecology of tanagers and honey-creepers in Trinidad. *Auk* **88**,291–322.

13 Vuilleumier, F. (1970) Insular biogeography in continental regions. I. The northern Andes of South America. *American Naturalist* **104**,373–388.

14 Vuilleumier, F. (1973) Insular biogeography in continental regions. II. Cave faunas from Tessin, southern Switzerland. *Systematic Zoology* **22**,64–76.

15 Brown, J.H. & Kodric-Brown, A. (1977) Turnover rates in insular biogeography: an effect of migration on extinction. *Ecology* **58**,445–449.

16 Janzen, D.H. (1968) Host plants as islands in evolutionary and contemporary time. *American Naturalist* **102**,592–595.

17 Janzen, D.H. (1973) Host plants as islands. II. Competition in evolutionary and contemporary time. *American Naturalist* **107**,786–790.

18 Bush, M.B. & Whittaker, R.J. (1991) Krakatau: colonization patterns and hierarchies. *Journal of Biogeography* **18**,341–356.

19 Simberloff, D.S. & Wilson, E.O. (1970) Experimental zoogeography of islands: a two-year record of colonization. *Ecology* **51**,934–937.

20 Simberloff, D.S. (1976) Species turnover and equilibrium island biogeography. *Science* **194**,572–578.

21 Schoener, T.W. & Spiller, D.A. (1987) High population persistence in a system with high turnover. *Nature* **330**,474–477.

22 Gilbert, F.S. (1980) The equilibrium theory of island biogeography: fact or fiction? *Journal of Biogeography* **7**,209–235.

23 Connor, E.F. & McCoy, E.D. (1979) The statistics and biology of the species-area relationship. *American Naturalist* **113**,791–833.

24 Diamond, J.M. (1969) Avifaunal equilibria and species turnover on the Channel Islands of California. *Proceedings of the National Academy of Sciences, USA* **64**,57–63.

25 Lynch, J.F. & Johnson, N.K. (1974) Turnover and equilibria in insular avifaunas, with special reference to the California Channel Islands. *Condor* **76**,370–384.

26 Diamond, J.M. (1971) Comparison of faunal equilibrium turnover rates on a

tropical island and a temperate island. *Proceedings of the National Academy of Sciences*, USA **68**,2742–2745.

27 Terborgh, J. & Faaborg, J. (1973) Turnover and ecological release in the avifauna of Mona Island, Puerto Rico. *Auk* **90**,759–779.

28 Jones, H.L. & Diamond, J.M. (1976) Short-time-base studies of turnover in breeding bird populations on the California Channel Islands. *Condor* **78**,526–549.

29 Diamond, J.M. & May, R.M. (1976) Island biogeography and the design of nature reserves. In *Theoretical Ecology: Principles and Applications* (ed. R.M. May). Blackwell Scientific Publications, Oxford, pp.163–186.

30 Simberloff, D.S. (1983) When is an island community in equilibrium? *Science* **220**,1275–1277.

31 Simpson, B.B. (1974) Glacial migrations of plants: island biogeographical evidence. *Science* **185**,698–700.

32 Diamond, J.M. (1982) Biogeographic kinetics: estimation of relaxation times for avifaunas of Southwest Pacific Islands. *Proceedings of the National Academy of Sciences*, USA **69**,3199–3203.

33 Simberloff, D.S. (1983) Biogeography: the unification and maturation of a theory. In *Perspectives in Ornithology* (eds A.H. Bush & G.A. Clark). Cambridge University Press, London, pp. 411–455.

34 Colwell, R.K. & Winkler, D.W. (1984) A null model for null models. *Ecological Communities and the Evidence* (eds D.R. Strong *et al.*). Princeton University Press, Princeton, pp. 344–359.

35 Shafer, C.L. (1990) *Nature Reserves. Island Theory and Conservation Practice.* Smithsonian Institution Press, Washington and London.

36 Western, D. & Ssemakula, S. (1981) The future of the savannah ecosystems: ecological islands or faunal enclaves? *South African Journal of Ecology* **19**,7–19.

37 Fryer, G. & Iles, T.D. (1972) *The Cichlid Fishes of the Great Lakes of Africa: their Biology and Evolution.* Oliver & Boyd, Edinburgh.

38 Carlquist, S. (1965) *Island Life.* Natural History Press, New York.

39 Carlquist, S. (1970) *Hawaii, a Natural History.* Natural History Press, New York.

40 Stone, B.S. (1967) A review of the endemic genera of Hawaiian plants. *Botanical Review* **33**,216–259.

41 Zimmerman, E.C. (1948) *The Insects of Hawaii. I. Introduction.* University of Hawaii Press, Honolulu.

42 Robertson, F.W. (1970) Evolutionary divergence in Hawaiian *Drosophila. Science Progress, Oxford*, **58**,525–538.

43 Carson, H.L. *et al.* (1970) The evolutionary biology of the Hawaiian Drosophilidae. In *Essays in Evolution and Genetics in Honour of Theodosius Dobzhansky* (eds M.K. Hecht & W.C. Steere). North-Holland, Amsterdam, pp. 437–543.

44 Carr, G.D. *et al.* (1989) Adaptive radiation of the Hawaiian silversword alliance (Compositae — Madiinae): a comparison with Hawaiian picture-winged Drosophila. In *Genetics, Speciation and the Founder Principle* (eds L.Y. Giddings, K.Y. Kaneshiro & W.W. Anderson). Oxford University Press, New York and Oxford, pp. 79–95.

45 Raikow, R.J. (1976) The origin and evolution of the Hawaiian honey-creepers (Drepanididae). *Living Bird* **15**,95–117.

46 Sibley, C.G. & Ahlquist, J.E. (1982) The relationships of the Hawaiian honeycreepers (Drepaninini) as indicated by DNA–DNA hybridization. *Auk* **99**,130–140.

47 Olson, S.L. & James, H.F. (1991) Descriptions of thirty-two new species of birds from the Hawaiian Islands: Part I. Non-Passeriformes. *Ornithological Monograph* **45**,1–88.

48 James, H.F. & Olson, S.L. (1991) Description of thirty-two new species of birds from the Hawaiian Islands: Part II. Passeriformes. *Ornithological Monographs* **46**, 1–88.

49 Briggs, J.C. (1974) *Marine Zoogeography*. McGraw-Hill, New York and London.
50 Keller, G., Zenker, C.E. & Stones, S.M. (1989) Late Neogene history of the Pacific Caribbean gateway. *Journal of South American Earth Sciences* **2**,73–108.
51 Ekman, S. (1953) *Zoogeography of the Sea*. Sidgwick & Jackson, London.
52 Grigg, R. & Hey, R. (1992) Paleoceanography of the tropical Eastern Pacific Ocean. *Science* **255**,172–178.
53 Fallaw, W.C. (1979) Trans-North Atlantic similarity among Mesozoic and Cenozoic invertebrates correlated with widening of the ocean basin. *Geology* **7**,398–400.
54 White, B.N. (1986) The isthmian link, antitropicality and American biogeography: distributional history of the Atherinopsidae (Pisces; Atherinidae). *Systematic Zoology* **35**,176–194.
55 Briggs, J.C. (1987) Antitropical distribution and evolution in the Indo-West Pacific Ocean. *Systematic Zoology* **36**,237–247.

7 *Patterns in the past*

As explained in Chapter 4, one way of grouping the terrestrial biological communities of the world is to place each of them in one of eight biomes. Each of these is distinguished from the others by its characteristic climate. A particular biome, such as desert, may therefore exist in many different parts of the world. In each desert live animals and plants of broadly similar appearance and way of life, but these may belong to quite different taxonomic groups from those found in a similar desert in another part of the world. A map showing the distribution of the biomes therefore tells us nothing about the patterns of distribution of taxonomic groups of animals or about the way in which different groups replace one another.

An alternative approach to classifying the patterns of distribution is to subdivide the world's surface into regions which appear to differ fundamentally from one another in the types of plant or animal to be found there. Although very few groups have precisely the same pattern of geographical distribution, there are some zones that mark the limits of distribution of many groups. This is because these zones are barrier regions, where conditions are so inhospitable to most organisms that few of them can live there. For terrestrial animals, any stretch of sea or ocean proves to be a barrier of this kind — except for flying animals whose distribution is for this reason obviously wider than that of solely terrestrial forms. Extremes of temperature, such as exist in deserts or in high mountains, constitute similar (though less effective) barriers to the spread of plants and animals.

These three types of barrier — oceans, mountain chains and large deserts — therefore provide the major discontinuities in the patterns of the spread of organisms around the world. Oceans completely surround Australia. They also virtually isolate South America and North America from each other and completely separate them from other continents. Seas, and the extensive deserts of North Africa and the Middle East, effectively isolate Africa from Eurasia. India and South-East Asia are similarly isolated from the rest of Asia by the vast, high Tibetan Plateau, of which the Himalayas are the southern fringe, together with the Asian deserts which lie to the north.

Each of these land areas, together with any nearby islands to which its fauna or flora has been able to spread, is therefore comparatively isolated. It is not surprising to find that the patterns of distribution of both the faunas (*faunal provinces* or *zoogeographical regions*) and the floras (*floral regions*) largely reflect this pattern of geographical barriers.

Before the detailed composition of these faunal provinces and floral regions can be understood fully, it is first necessary to follow the ways in which today's patterns of geography, climate and distribution of life came into existence. From what has been discussed in earlier chapters, it is clear that the differences between the faunas and floras of different areas might be due to a number of factors. Firstly, any new group of organisms will appear first in one particular area. If it competes with another, previously established group in that area, the expansion in the range of distribution of the new group may be accompanied by contraction in that of the old. However, once it has spread to the limits of its province or region, whether or not it is able to spread into the next will depend initially on whether it is able to surmount the geographical ocean or mountain barrier, or to adapt to the different climatic conditions to be found there. (Though, even if it is able to cross to the next province or region, it may be unable to establish itself because of the presence there of another group which is better adapted to that particular environment.) Of course, changes in the climate or geographical pattern could lead to changes in the patterns of distribution of life. For example, gradual climatic changes, affecting the whole world, could cause the gradual northward or southward migrations of floras and faunas, because these extended into newly favourable areas and died out in areas where the climate was no longer hospitable. Similarly, the possibilities of migration between different areas could change if vital links between them became broken by the appearance of new barriers, or if new links appeared.

Plate tectonics

Overwhelmingly the most important factor in causing major, long-term changes in the patterns of distribution of organisms has been the slow alteration in the geography of the world that used to be called continental drift, but is now known as plate tectonics. This not only affects the distributional patterns directly, by the splitting and collision of land masses and their movement across the latitudinal bands of climate; it also affects them indirectly, as new mountains, oceans or land barriers deflect the atmospheric and oceanic circulations, so changing the climatic patterns upon the land masses.

The motive force for plate tectonics is sea-floor spreading, caused by great convection currents that bring heated material to the surface from the hot interior of the earth. Where these upward currents reach the surface at the floors of the oceans, chains of underwater volcanic mountains form. These are known as 'spreading ridges', because new sea floor is formed there and is moved progressively away from the spreading ridge as more new material is steadily produced there. The age of the ocean floor is therefore progressively greater as one moves, to one side or the other, further away from the spreading ridges. Old ocean floor is consumed at the deep troughs or 'trenches' around the edges of the Pacific Ocean, where it disappears back into the earth (Fig. 7.1). Where these lie adjacent to the edge of a continent, the material of which is too light to be drawn back into the earth, the volcanism caused by the inwards movement of old ocean floor raises mountain chains such as the Andes of today.

Where spreading ridges lie within the oceans, their activity will cause continents to move apart, and ultimately may cause them to collide with one another, the collision raising mountains such as the Himalayas and Urals. A ridge may also gradually elongate and extend under a continent; its activity will then cause the gradual rifting apart of those regions of the continent that lie on either side of the spreading ridge. As these move apart, they become separated by a new, widening ocean, the floor of which is similarly moving to one side or the other,

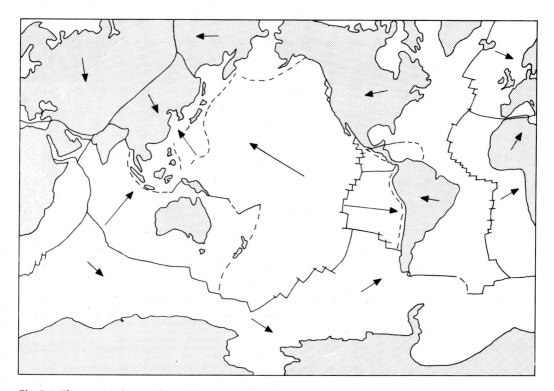

Fig. 7.1 The major tectonic plates today. Lines within the oceans show the positions of spreading ridges: dotted lines show the positions of trenches. Lines within the continents show the divisions between the Eurasian, African, Indian, China and North American plates. Arrows show the directions and proportionate speeds of movement of the plates. The Antarctic plate is moving clockwise.

away from the spreading ridge. The moving units at the surface of the earth are therefore areas that may contain continental masses, or that may consist only of ocean floor. These units are known as 'tectonic plates', and earth scientists today refer to 'plate tectonics' rather than to 'continental drift'.

The movements of the continents are quite slow — only about 5–10 cm per year. These movements must have affected life in several ways, even though the changes must have been incredibly gradual, and noticeable only over a period of millions of years. The most obvious change would have resulted directly from the movements of the continents relative to the poles of the earth, and to the Equator. As they moved, so the different areas of land would have come to lie in cold polar regions, in cool, damp temperate regions, in dry subtropical regions, or in the hot, wet equatorial regions.

The true edges of the continents are marked, not by their coastlines, but by the edge of the continental shelf. Between the continental shelves the deep oceans separate the continental plates. Sometimes the whole of these plates has been above sea level. At other times comparatively shallow 'epicontinental' seas have covered the edges of the continents (for example the North Sea today) or formed seas within the continents (such as Hudson's Bay today). Though the extents of these shallow seas have varied over geologic time, they must have formed a barrier to the spread of organisms, and they are shown in the palaeogeographic maps (Figs 7.3, 7.5). They were particularly extensive in the Jurassic and the Cretaceous Periods of geological time (Fig. 7.2).

Drift also affected the climates of the continents in other, less direct ways. The climate of any area is controlled by its distance from the sea, which is the ultimate source of rainfall. The central part of great super-continental land masses, such as Eurasia today or Gondwana in the past, is therefore inevitably dry, and the climate of such areas in monsoonal (see p. 81), experiencing great daily and seasonal changes of temperature. The break-up of a supercontinent, or the spread of shallow epicontinental seas into the interior of continents, would have brought moister, less extreme climates to these regions. The extent of these seas may also have been at least partly the result of plate tectonic activity. There appears to be a correlation between the total length of the system of oceanic spreading ridges and the extent of the transgression of seas over the continents. Because the ridges are formed of high underwater mountain chains, an increase in their length will decrease the capacity of the ocean basins and cause an increase in the areas of the continents that are covered by epicontinental seas.

The pattern of ocean spreading ridges, as well as the resulting splitting and movement of the continents themselves, will also have affected their climates by altering the patterns of water circulation in the oceans.

Era	Period	Epoch	Approximate duration in millions of years	Approximate date of commencement in millions of years before present	
Cenozoic	Quaternary	Pleistocene	2·4	2·4	
Cenozoic	Tertiary	Pliocene	2·6	5	Millions of years ago
Cenozoic	Tertiary	Miocene	18	23	
Cenozoic	Tertiary	Oligocene	13·5	36·5	50
Cenozoic	Tertiary	Eocene	16·5	53	
Cenozoic	Tertiary	Paleocene	12	65	
Mesozoic	Cretaceous		70		100
Mesozoic	Jurassic		70	135	150
Mesozoic	Triassic		45	205	200
Palaeozoic	Permian		40	250	250
Palaeozoic	Carboniferous		65	290	300
Palaeozoic	Devonian		50	355	350
Palaeozoic	Silurian		25	410	400
Palaeozoic	Ordovician		75	435	450
Palaeozoic	Cambrian		60	510	500
Proterozoic			4000	570	550
					4600

Formation of Earth's crust about 4600 million years ago

Fig. 7.2 The geological time scale.

For example, the cold Humboldt Current up the western side of southern Africa, the warm clockwise Gulf Stream of the North Atlantic, and the southward movement of cold deep water from the Arctic Ocean into the North Atlantic, all result from the present pattern of continents, and all affect the climates of the adjacent continents. The different continental patterns of the past would therefore have resulted in different patterns of ocean circulation, and different climates.

Furthermore, the distribution of climate within the continental masses must also have been affected by the appearance of new mountain ranges as a result of continental drift. These would have had particularly great effects on the climate of the continents if they arose across the paths of the prevailing moisture-bearing winds, since areas in the lee of the mountains would then become desert. These can be seen today in the Andes, to the east of the mountain chain in southern Argentina, and to the west along the coast from northern Argentina to Peru; the winds in these two regions blow in opposite directions. A huge area of desert, including the arid wastes of the Gobi Desert of outer Mongolia, has also formed in central Asia, far from seas from which winds could gain moisture to fall as rain.

The evidence for past geographies

The simple pattern of age-banding of the ocean floor that results from the process of sea floor spreading provides direct evidence of the positions of the continents over the last 180 million years, to the middle of the Jurassic Period. By removing from the map any ocean floor younger than, for example, the end of the Cretaceous Period, one can return the continents to their positions at that time. But all ocean floor older than 180 million years has been consumed, and the positions of the continents before that time therefore have to be deduced from a different line of evidence — palaeomagnetism. This relies on the fact that many rocks contain particles of iron-containing minerals. As the rock cooled after its deposition, these particles became aligned along the lines of the then-prevailing direction of the earth's magnetic field, like tiny compass needles. It is therefore possible to calculate how far from the magnetic pole the rocks lay when they were deposited, and in which direction the pole lay. This indicates the orientations of the continents and their north–south positions (but not their east–west, longitudinal positions). Other evidence for the past positions of the continents can be gained from the types of rocks that were laid down within them at that time (e.g. desert sandstones or glacial deposits). The former patterns of union of continents can also be deduced from matching the sequences of rock types or rock ages, and from dating the times of the rise of mountain chains that mark their collision.

But the evidence for past geographies does not only come from physical geology, for palaeobiogeography has always played an important role, and was particularly important in the beginnings of the formulation of the theory of continental drift. The realization that the distributions of Permo-Carboniferous plants in the Southern Hemisphere, and of various Mesozoic reptiles on either side of the South Atlantic, did not make simple sense in the context of modern geography helped to convince the German scientist Arthur Wegener that the continents must have moved [1]. Many palaeobiogeographers believed in the correctness of Wegener's theory (first published in 1912) long before the new sea floor and palaeomagnetic data of the 1960s convinced the geologists. Nevertheless, the degree of detail provided by the geophysical data is now on the whole greater than that of the palaeobiogeographical evidence, so that the latter more often has only a confirmatory role. The geophysical data, by establishing the times at which land masses split or united, have also identified the units of time and geography within which it is appropriate to make palaeobiogeographic analyses [2]. Until then, such analyses often made little sense, for they frequently combined units of time within which major changes of geography had taken place, or combined geographical areas in inappropriate patterns.

Today, we can see that patterns of distribution of plants in the Permo-Carboniferous, and of vertebrates in the Cretaceous conform closely to the patterns of land and sea that are deduced by geologists when the continents are placed in the locations suggested by the geophysical data, and the outlines of shallow seas are added (see below). Fossil marine faunas provide evidence on the development of faunal provinces on either side of widening oceans (p. 165). The shapes of the fossil leaves of flowering plants indicate the climatic regime of the areas they inhabited, as do the types of plant themselves (see below), and their pollen also provides information on climatic change during the Ice Ages (p. 258).

But although palaeobiogeography therefore often plays a subordinate role, there are also situations in which its data provide more direct, detailed evidence than the geophysical record. For example, both the fossil marine faunas on either side of the Panama isthmus (p. 164), and the fossil mammals of North and South America (p. 221), provide clearer and more reliable evidence of the date of the final linking of those two continents than does any geophysical data from the area. Because sea floor spreading does not provide data on the presence or spread of epicontinental seas, palaeobiogeography also provides crucial evidence on the times during which these seas separated areas of land.

Changing patterns of continents

A major feature of the geography of the late Palaeozoic and the Mesozoic (Fig. 7.2) was the great supercontinent we call Gondwana. It has long been known that this included the land areas that later became South America, Africa, Antarctica, Australia and India. More recently, it has been realized that its northern edge included an area that gradually broke up into a whole series of minor land masses, which geologists call 'terranes'. These moved northwards and joined the southern edge of Eurasia to form southern and south-western Europe, Tibet and two separate portions of China.

By the Silurian, 435–410 million years ago, when complex living organisms first colonized the land, North America and northern Europe had formed a second major land mass, Euramerica, while Siberia made up a third continent (Fig. 7.3a). Euramerica and Gondwana joined together in the Late Carboniferous to Early Permian, 300–270 million years ago (Fig. 7.3b). Siberia joined this enormous land mass in the Late Permian, 260 million years ago, and the two portions of China joined with this in the Triassic, to form the single world continent we call Pangaea (Fig. 7.3c). Throughout this period of time, Gondwana had gradually been moving across the South Pole.

However, it was not long before Pangaea started to become divided. Gondwana separated from the more northern land mass, which comprised North America and Eurasia, and which is called Laurasia. While Gondwana started to break up into separate continents, shallow epicontinental seas penetrated across Laurasia. Both these processes started in the Jurassic and continued throughout the Cretaceous Period (Fig. 7.5). In the Late Cretaceous (Fig. 7.5c) Europe was still separated from Asia by a sea known as the Obik Sea, and the Mid-Continental Seaway completely bisected North America into eastern and western land areas. The western part was connected to Asia via Alaska and Siberia to form a single 'Asiamerican' land mass, and the eastern portion was connected to Europe via Greenland to form a 'Euramerican' land mass. At the same time, South America finally became separated from Africa, but it may have remained connected to Antarctica (and thence also to Australia) until well after the end of the Cretaceous (see below). India separated from the rest of Gondwana in the Early Cretaceous.

It was not until the Early Cenozoic that the geography of the world became similar to that we see today (Fig. 7.5d). In the Eocene, Australia finally separated from Antarctica, and India became united with Asia. At about the same time, Europe became separated from Greenland, and the drying of the Obik Sea finally made Europe continuous with Asia. Though the continents of Africa and Eurasia had for long been close together, shallow seas separated their land areas from one another until the Miocene. The final link between South America and North America did not form until the Late Pliocene, about two million years ago.

Early land life on the moving continents

Our understanding of the movements of the continents and of the timing of the different episodes of continental fragmentation or union is

(a)

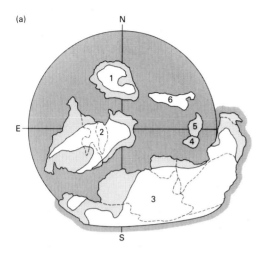

(b)

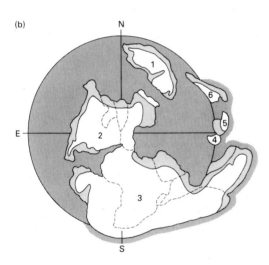

(c)

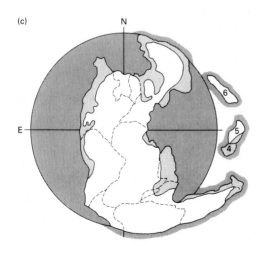

Fig. 7.3 World geography in (a) the Early Devonian;
(b) the Late Carboniferous; (c) the Late Permian.
Lambert equal-area projection: areas lying within the
circle lie on the front hemisphere of this view of the
globe, and any areas that are outside this circle lie on
the back hemisphere. Some of the areas along the
northern margin of Gondwana, that later contributed
to southern Asia, have been omitted for simplicity and
clarity. 1, Siberia; 2, Euramerica; 3, Gondwana; 4,
South-East Asia; 5, southern China; 5, northern China.
E, Equator; N, North Pole; S, South Pole. Dark tint
indicates ocean, light tint indicates epicontinental
seas. Dotted lines indicate the future coast-lines of the
modern continents. (Continental outlines and
positions from Scotese and McKerrow [3]; outlines of
land areas from Stanley [4].

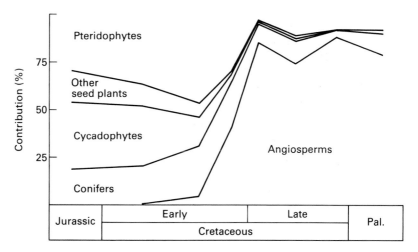

Fig. 7.4 The percentage contribution of the major plant groups to Jurassic Cretaceous and Paleocene (Pal) floras. Pteridophytes include ferns and lycopods; cycadophytes include cycads, bennettitaleans and pinnate-leaved seed ferns. After Crane [24]. Copyright 1989 by the AAAS.

now fairly detailed, and the distribution of fossil organisms correlates very well with the varying patterns of land.

The earliest time at which there is enough evidence to discern patterns of life is the Early Devonian, 380 million years ago (see Fig. 7.3a), by which time separate floras and fish faunas can be distinguished in the cool, northerly placed Siberian continent, the equatorially placed Euramerican continent and the eastern region of Gondwana [5,6]. Land vertebrates, the amphibians, evolved during the Late Devonian, and the first reptiles evolved from them soon afterwards. The fact that, apart from members of an early and divergent amphibian group known as the ichthyostegids, all the early fossils of these two groups are found only in Euramerica, suggests that they may have evolved on that continent [2,7]. That would not be surprising, for Euramerica then lay in a near-equatorial position, which would have encouraged a rich growth both of plants and of the invertebrates that the earliest terrestrial vertebrates would have fed upon.

The great expansion of the land plants began in the Devonian, and this may well itself have had an effect on the world's climate. Today, vegetated surfaces decrease the albedo (or reflectivity) of an area by 10–15 per cent, while plants recycle much of the rainfall (up to 50 per cent in the Amazon basin) [8]. Furthermore, the photosynthetic activities of the plants would have reduced the CO_2 content of the atmosphere, which would have lowered the temperature. The development of the earth's vegetation, and these consequential effects, would have taken many millions of years, and it is tempting to speculate as to whether there might be a connection with the next significant event, which was the onset of global cooling.

This cooling began in the middle of the Carboniferous Period, and led to the appearance of ice sheets around the South Pole, similar to those of Antarctica today. As Gondwana moved across the South Pole, the glaciated area moved across its surface. Though the whole South Polar area must have been icy cold, the proportion that was actually glaciated probably varied according to the position of the Pole itself. When this was near to the edge of the supercontinent, the adjacent ocean would have provided enough moisture to create the heavy snow-falls that would have formed extensive continental ice sheets. But when the Pole lay further inland, away from the ocean, it is likely that the inland areas, although cold, would not have been heavily glaciated.

This polar glaciation caused the latitudinal ranges of the Carboniferous floras to be compressed towards the Equator. In the equatorial region there was a great swampy tropical rain forest, rather like that of the Amazon Basin today. This lay across Euramerica and was fed by rains from the warm westward equatorial ocean current that would have washed the eastern shore of that continent. Surprisingly, the distribution of different types of rock laid down at this time shows no sign of the presence of monsoonal, dry conditions in the interior of the great super-continent, and the climatic pattern seems to have been essentially latitudinal [9]. The equatorial wet belt was bordered to the south by a subtropical desert that stretched across northern South America and northern Africa; beyond this lay the cold lands around the glaciated South Pole. Another desert belt covered northern North America and north-eastern Europe, but Siberia lay in a wetter zone further to the north.

The absence of dormant buds and of annual growth rings in the fossil remains of the equatorial coal-swamp flora indicates that it grew in an unvarying, seasonless climate. The flora was dominated by great trees belonging to several quite distinct groups. *Lepidodendron*, 40 m tall, and *Sigillaria*, 30 m tall, were enormous types of lycopod (related to the tiny living club-moss *Lycopodium*). Equally tall *Cordaites* was a member of the group from which the conifer trees evolved, and *Calamites*, up to 15 m high, was a sphenopsid, related to the living horse-tail, *Equisetum*. Tree-ferns such as *Psaronius* grew up to 12 m? high, and seed-ferns such as *Neuropteris* were among the most common smaller plants living around the trunks of all these great trees. In the eastern United States, and in parts of Britain and central Europe, the land covered by this swamp forest was gradually sinking. As it sank, the basins that formed became filled with the accumulated remains of these ancient trees. Compressed by the overlying sediments, dried and hardened, the plant remains have become the coal deposits of these regions.

Far to the south of the equatorial coal-swamp flora, a very different flora covered the lands around the growing South Polar ice sheets. It is known as the *Glossopteris* flora, after a genus of woody shrub that is found only in that region. *Calamites* trees are absent from this flora, and lycopods, ferns and seed-ferns were less important than in the north. (Surprisingly, the *Glossopteris* flora is found as close as 5° to the then South Pole, in a region that must have been not only cold but also with a very short winter day-length. The very marked seasonal growth rings of plants from these regions are therefore not surprising, though the size of these is unexpectedly great from so high a latitude.)

After the Carboniferous gave way to the Permian, the coal swamps of southern Euramerica disappeared and, by the Late Permian, deserts lay in their place. This was partly because these regions had moved northwards, away from the Equator, and partly because the mountain ranges of northern Africa and eastern North America had extended and risen higher, blocking the moist winds from the ocean that lay to the east.

Three different Permian floras can be distinguished [10]. To the north of the northern subtropical desert lay the temperate 'Angara' flora, with *Cordaites*-like conifers and herbaceous horsetail plants, ferns and seed-ferns. The flora is richest towards the eastern coast, and becomes less diverse towards the colder north of Siberia. The second flora is the rich, varied, tropical 'Cathaysian' rain-forest flora, with sphenopsid and *Lepidodendron* trees, *Gigantopteris* lianas and many types of seed-fern. Conifers and *Cordaites* were rare in this flora, which is found along the eastern margins of Euramerica and in the land masses that were moving from Gondwana towards Asia, across the tropical ocean. The third flora is the south-temperate Gondwana flora, descendant from the earlier Late Carboniferous *Glossopteris* flora of that continent described above.

The land vertebrates that had evolved in Euramerica in the Carboniferous did not at first colonize the rest of the world [2]. Their dispersal southwards into Gondwana may have been hindered by several factors. Mountains lay along the south-eastern margin of North America, where this region had collided with Gondwana, while the deserts further south must have added an additional barrier to dispersal. This may be the reason why land vertebrates are unknown from Gondwana until the Late Permian, after the disappearance of the polar ice cap. Not surprisingly, land vertebrates did not reach Siberia or China until after those land masses had joined the world supercontinent in the Late Permian.

In the Late Permian, as in the Late Carboniferous, evidence for the presence of arid areas in the centre of the mid-latitudinal regions of the supercontinent is lacking. In fact, rich faunas of Late Permian fossil reptiles have been found in regions of southern South America and Africa that, according to climatic modelling experiments, would have had annual temperature changes of 40–50°C — similar to those of Central Asia today, and not very congenial to reptiles [11].

One world — for a while

The coalescence of the different continental fragments to form Pangaea had geographical effects and also consequential climatic effects. The disappearance of the oceans that had previously separated the continents, and the formation of the enormous supercontinent, had left vast tracts of land now far from the oceans and the moist winds that originated there. Furthermore, the new, lofty mountain ranges that still marked the regions where Euramerica, Siberia and China had collided, provided physical and climatic barriers to the dispersal of their faunas and floras. Though one would in any case have expected the Permian floras described above to change through time, these barriers made it more likely that any changes would take place independently in each, rather than these floras gradually blending. So we can identify a general evolutionary change, in which older types of tree, such as those belonging to the lycopods and sphenopsids, and *Calamites*, disappeared. They were replaced either by the radiation of existing types of tree, such as caytonias and ginkgos, or by new groups such as cycads and bennettitaleans. A new type of fern, the osmundas, also appeared. This floral change was complete by the end of the Triassic Period. In addition, we can also identify changes within the floras. This is most clearly seen in the Gondwana flora, where in the Early Triassic *Glossopteris* itself was replaced by the seed-fern *Dicroidium*.

In the Jurassic and Early Cretaceous, floras seem to have gradually become more similar to one another, approaching the modern pattern in which there are gradual latitudinal changes governed by climate, and these manifest themselves as changing patterns in the dominance of different groups as one moves from lower latitudes to higher latitudes. For example, in the Jurassic and Early Cretaceous, two floral provinces can be distinguished in the northern land mass, Laurasia [12] (Fig. 7.5a). The Northern Laurasian (or 'Siberian–Canadian') province has a lower floral diversity, with more deciduous plants such as conifers, and with evidence of annual growth rings, while the Southern Laurasian (or 'Indo-European') province does not show these features. From the mid-Jurassic onwards there was a floral change that was more marked in the Southern Laurasian province, and which may have been linked to increasing aridity. Several types of fern and of sphenopsid became extinct, and the cycads became rare. Among the conifers the Cheirolepidiaciae became important, and the araucarias were numerous in the moister, low to middle latitudes.

The southern boundary between the Northern and the Southern Laurasian floral provinces in Asia shifted northwards between the Late Jurassic and the Early Cretaceous. This seems to have been caused by a climatic change, the southern part of Asia becoming increasingly hot and arid. The change is sometimes quoted as evidence for a general increase in global warmth. However, there is no evidence for this elsewhere, and it seems instead to reflect a more local phenomenon, perhaps related to changes in the temperatures of the ocean currents adjacent to southern Asia, and therefore to the rainfall of this area. These Asian Mesozoic floras extended to high (about 70°) northern and southern latitudes, to areas that, although clearly warm, must have had seasonal very brief periods of daylight [13].

To outline now the biogeographic history of the vertebrate animals (amphibians and reptiles) of Pangaea, one must return to the Permian and Triassic. By the middle of the Permian, these animals appear to have been quite competent at dispersing through regions of different climate, and Pangaea soon came to contain a fairly uniform fauna, with little sign of distinct faunal regions [14]. Nearly 60 families of Triassic land vertebrate have been described, and their distribution provides overwhelming evidence for the existence of a Pangaea world continent within which they could travel freely. The Triassic faunas of most of today's continents can be compared with one another (Table 7.1). (Those of Antarctica and Australia are still too poorly known to be analysed in this way, but what has been found shows clearly that the same Triassic

Table 7.1 Coefficients of faunal similarity at family level between the Triassic faunas of today's continents. The number shown in parentheses after each continent shows the number of terrestrial Triassic families found in that continent; where the presence of a family is not yet fully confirmed, a score of 0·5 has been added.

	Europe (41)	Asia (20)	South America (20)	Africa (42·5)	India (17)
North America (21)	71%	40%	55%	71%	59%
	Europe	85%	70%	61%	88%
		Asia	47·5%	90%	44%
			South America	70%	50%
				Africa	81%

fauna was present in those two areas also.) The comparison uses the coefficient of faunal similarity, $100\,C/N_1$, in which C is the number of families common to the two continents being compared, and N_1 is the number of families in whichever of those two has the smaller fauna. Thus, for example, 15 of the 21 families found in the Triassic of North America are also known in Europe, and the coefficient for this comparison is therefore $100 \times 15/21 = 71$ per cent. As can be seen from Table 7.1, these faunas are remarkably similar to one another — in fact, no less similar than the mammal faunas of New York State and Oregon today.

Great changes took place in the world-wide fauna during the Triassic [15]. The bulk of the Permian faunas had been made up of mammal-like reptiles and other older types of reptile. These disappeared during the Early and Middle Triassic, and were at first replaced by a radiation of early reptiles known as archosaurs. These in turn were soon replaced (in the Late Triassic) by their own descendants, the dinosaurs, which came to dominate the world throughout the Jurassic and Cretaceous. Though comparatively little is known about the Jurassic dinosaurs, it is enough to show that they were able to spread throughout the world. Their route between North America and Asia must have been via Alaska and Siberia, which still had quite mild climates (see below). That the dinosaurs were also able to reach Gondwana is shown by the similarities between the Jurassic dinosaur faunas of North America and East Africa. In these two areas are found not only the same families of dinosaur, but also in some cases the same genera — for example, the sauropod dinosaurs *Brachiosaurus*, *Bothriospondylus* and *Barosaurus*, and the ornithopod *Dryosaurus*. Although the position of the land connection between Gondwana and the north is unknown, it seems likely that it was via South America.

The distribution of the new types of dinosaur that evolved during the Cretaceous mirrors the palaeogeographic maps very closely. Most of those that evolved during the Early Cretaceous (ostrich-dinosaurs, dome-headed dinosaurs, dromaeosaurs and primitive duck-billed dinosaurs) dispersed throughout the Northern Hemisphere, which was still undivided by seaways (Fig. 7.5a). Those types that evolved later, near the end of the Cretaceous, however, showed a more restricted pattern of distribution: tyrannosaurs, protoceratopsids and advanced duck-billed dinosaurs are all found only in Asiamerica, the new north–south seaways having prevented them from reaching Euramerica [2] (Fig. 7.5c). The rhinoceros-like ceratopsians, which only evolved in the very latest Cretaceous in North America, did not even reach Asia, which suggests that a marine barrier had by then separated the two continents.

There is little evidence for the Late Cretaceous groups in Gondwana, though this is partly because much less has so far been discovered of the Cretaceous record of the Southern Hemisphere. In the Late Cretaceous, duck-billed dinosaurs, sauropods and a possible ceratopsian are known from South America. There must, therefore, have been some dispersal route between North and South America during at least part of the Cretaceous, and this is supported by the Late Cretaceous dispersal of some mammals between the two continents (see below). Since there was no wholesale exchange of their faunas, it must have been either a narrow, Panama-like land-bridge, or a sweepstakes route similar to the West Indies island chain of today.

The early spread of mammals

The first, most primitive, mammals appeared in the Triassic Period, not long after the first dinosaurs, but they almost certainly laid eggs, as do the living monotremes (the platypus and spiny anteater) of Australia. The two major modern groups, the marsupials and the placentals, did not appear until the Late Cretaceous. In the marsupial type of mammal the young leave the uterus at a very early stage and then complete their development in the mother's pouch, whereas in the placental mammals the whole period of embryonic development takes place in the uterus. Though the placentals spread to all parts of the world except Australia (see below), the marsupials were only successful in the Southern Hemisphere. Unfortunately, our understanding of the patterns of distribution of these two groups is hampered by our incomplete knowledge of the timing of some crucial plate-tectonic events and of the composition of some early faunas.

The sea-floor spreading record provides reliable data on the timing of some plate-tectonic events. For example, we know that the last land contact between South America and Africa finally broke at the end of the Early Cretaceous, about 110 million years ago. Similarly, though sea-floor spreading between Antarctica and Australia commenced in the Early Cretaceous, final separation between the two continents was not until the Late Eocene, 40 million years ago [16]. India had parted company from Gondwana much earlier, in the Early Cretaceous, 118 million years ago. In some areas, however, complex plate-tectonic events were taking place and clear sea-floor spreading data are absent. This includes the regions that link South America northwards to North America and southwards to Antarctica, so that the exact timing of the separations of these continents is uncertain, though they seem to have parted some time in the Early Cenozoic. The fossil record, too, is exasperatingly silent on the mammal faunas of Australia before the Oligocene, while little is known of the Cretaceous floras of India. As a result of these inadequacies in our knowledge, we have to piece together a plausible history of the plate-tectonic and biogeographic history through this time.

(a)

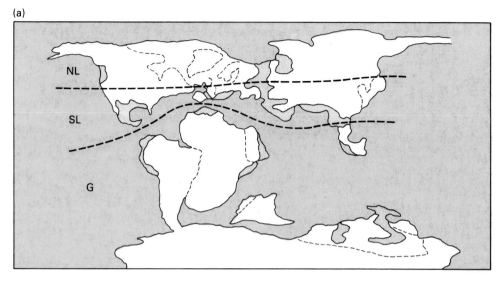

(b)

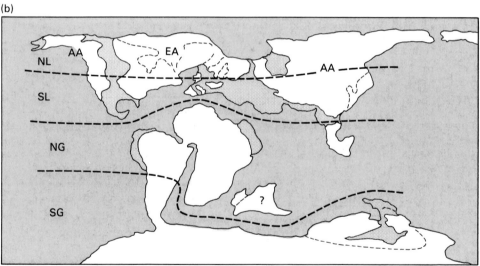

Fig. 7.5 World geography in (a) the Early Cretaceous; (b) the Middle Cretaceous; (c) the Late Cretaceous; (d), the Late Eocene. Mercator projection; after Barron [30]. Light shading indicates ocean; dark shading indicates epicontinental seas. Dotted lines indicate modern coastlines. Floral provinces in (a)–(c) (after Crane [27]): A, *Aquilapollenites*; G, Gondwanan; N, Normapolles; NG, Northern Gondwanan; NL, Northern Laurasian; SG, Southern Gondwanan; SL, Southern Laurasian and in (b) AA, Asiamerica; EA, Euramerica.

The earliest placentals are known from the Late Cretaceous of Asia and some of them, like the contemporary Asian dinosaurs, spread to western North America. The earliest marsupials, on the other hand, are known from the Late Cretaceous of both North and South America; they are also known as fossils in Eocene deposits in Antarctica [17]. The

(c)

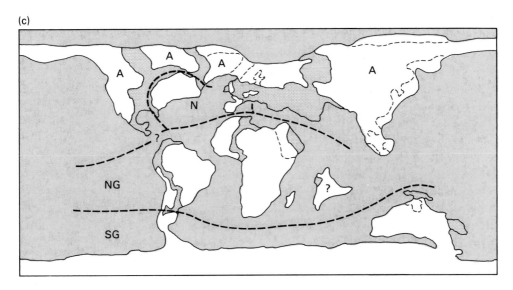

(d)

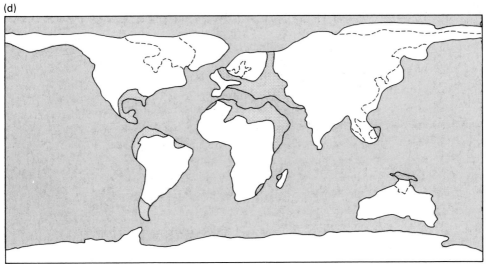

Fig. 7.5 (Continued)

simplest explanation of the final pattern of distribution of the two groups is to assume that marsupials evolved somewhere in the South America–Antarctica–Australia chain of continents (Fig. 7.5c,d). They then dispersed through all three of those continents and also northwards to North America, presumably over the same route used by the Late Cretaceous invasion of dinosaurs from North America to South America, and reached Europe via Greenland (see p. 230). In North America these marsupials met placentals that had dispersed to that continent from Asia. Some of these placentals then continued southwards to South America, where they coexisted with the marsupials — but neither of these groups were able to reach Africa or India, for both those continents had already separated from the rest of Gondwana.

Until recently, monotremes (either living or fossil) were unknown outside Australia. However, these egg-laying mammals are known as far back as the Early Cretaceous, at a time when South America, Antarctica and Australia were still connected to one another, and when Antarctica was still warm and forested. It was therefore always likely that monotremes had once inhabited all three of those continents, and their recent discovery in the early Paleocene of South America [18] is not surprising and consequently has not raised any new palaeobiogeographic problems.

A second aspect of the biogeography of early mammals has, however, been transformed by the recent report of a fossil placental mammal in Australia. Until then it was necessary to suggest some theoretical reason for the fact that placentals (either fossil or living) were unknown from Australia — except for rats and bats, which had arrived in the Pliocene or Pleistocene. So, one could speculate that Australia (or Australia/ Antarctica) had separated from South America towards the end of the Cretaceous, before placentals had dispersed down through South America from their origin in Asia. But it then became clear that Australia had been connected to Antarctica (and thence to South America) as late as the Eocene. It was consequently necessary to suggest that there might have been some other geographical barrier that had prevented the placentals from dispersing from South America to Australia — perhaps a marine barrier within Antarctica or southern South America. The recent discovery of the tooth of what appears to be a South American type of placental in the early Eocene of Australia [19] makes all these geographical speculations unnecessary, but instead provides a zoological puzzle. In other parts of the world, marsupials seem to have been unable to withstand the competition of placentals. In the Early Cenozoic of the Northern Hemisphere, as in the Late Cenozoic of South America (see pp. 221–2), the arrival of placentals led to the elimination, or near-elimination, of the marsupials. Why, then, did the Early Cenozoic placentals of Australia become extinct, leaving the monotremes and the marsupials as the only Late Cenozoic mammals of that continent? Perhaps they were less able to adapt to the later climatic changes of Australia, with its increasing aridity (see p. 217), than the marsupials, the development of whose pouch-young can easily be terminated in unfavourable conditions.

There is, however, perhaps a moral to be learned from these two stories. It is inevitable that biogeographers should attempt to explain the palaeobiogeographical facts as they are known at any point in time. But we must also be aware of the inequality in our knowledge of the fossil record in the two Hemispheres. The northern record had been investigated by many more scientists and over a much greater period of time than that of the Southern Hemisphere. The absence of evidence that a particular group was present in the south cannot therefore be taken as strong evidence that it was absent from that area. It is very likely that further surprises await us as the Gondwana records of the dinosaurs of the Late Mesozoic and of the mammals of the Early Cenozoic are gradually discovered.

The great Cretaceous extinction event

The diversification of both the marsupials and the placentals was only made possible by the sudden disappearance of the dinosaurs, which had dominated the world since the Late Triassic, over 140 million years before. But the dinosaurs were not the only group to disappear at that time, for they were accompanied into extinction by other reptile groups, such as the flying pterosaurs and the great marine reptiles such as the plesiosaurs and mosasaurs. The ocean biota also lost the ammonoids, which had been in existence for 250 million years, as well as many types of marine plankton and of such invertebrate groups as reef-building corals and the bivalved, mollusc-like brachiopods. It has been estimated that 75 per cent of all the species, and 25 per cent of all families, of animal became extinct at that time. In some groups, such as the plankton and brachiopods, there was a later wave of replacement by new types, but other groups had disappeared for ever. What could have caused the simultaneous extinction of vertebrates and invertebrates (but not all of either group) on land, in the air and in the oceans?

The fact that many types of animal and plant became extinct at the boundary between the Cretaceous Period and the succeeding Tertiary Period (the 'K/T' boundary), had long been recognized, but there was also a prolonged debate as to whether the extinction was gradual or abrupt. It now seems that it was both. Detailed study of fossil-bearing strata across the K/T boundary suggests that, in many groups, there was a gradual reduction in their diversity towards the end of the Late Cretaceous. The cause of this is very likely to have been the climatic change that was taking place during this time (see below). But there does seem also, at the extreme end of the Cretaceous, to have been a final event, the duration of which is below our powers of resolution 65 million years later, and during which a very large number of taxa became extinct.

Though many explanations have been put forward as to the nature of this sudden event, more and more evidence is now accumulating that it was the result of the impact of an enormous meteor or 'bolide', about 10 km in diameter. The first evidence of this was the discovery by father and son Walter and Luis Alvarez, of Berkeley University, California, of unusually high levels of 'rare-earth' minerals, such as iridium and osmium, in a thin layer of clay laid down at the K/T boundary at Gubbio in Italy [20]. Significantly, these minerals were also present in precisely the proportions in which they are found in some types of meteorite. Later research has shown that this thin clay layer, together with the minerals, is found in over 100 locations as far apart as North America, New Zealand, the South Atlantic and the North Pacific, so it was clearly deposited world-wide, rather than being merely a local event. The Alvarez' suggestion that these deposits were the result of the impact of an enormous meteorite was later supported by the discovery in the clay layer of two other substances that are characteristic of meteorite impact — glassy 'microtektites' and shock-metamorphosed quartz grains [21]. The pattern of distribution of microtektites and other sediments derived from the bolide impact has helped to locate the position of the meteorite impact. Many facts now strongly suggest that the position is marked by the 180 km diameter crater, made 64.5 million years ago, at Chicxulub in northern Yucatan, Mexico.

The shock wave from the meteorite impact would have had a devastating influence on the forests and larger animals over a very wide area. A fireball, many hundreds of kilometres across, would have formed, and burning hot winds would have swept around the world. This, together with the red-hot ejecta from the crater, would have started forest fires in many areas. The heat would have produced quantities of nitric acid in the atmosphere, which would have produced acid rain, defoliated many plants, and reduced the ozone content of the atmosphere and so allowed more of the damaging ultraviolet radiation to penetrate to the earth's surface. The K/T boundary clay also contains large quantities of soot — so much that it has been suggested that 90 per cent of the world's forests may have burned, releasing quantities of pyrotoxins into the atmosphere.

Studies of the sediments at the Chicxulub crater suggest that between 1000 and 4000 cubic kilometres of rock were thrown up by the impact. The minerals of the rocks where the meteorite struck contained carbonates and sulphates; their heating in the impact would have released great quantities of CO_2 and SO_2 into the atmosphere, as well as water vapour from the heated ocean. These would have had profound and complicated effects upon the world's weather [22].

The SO_2 and water vapour would have combined in the atmosphere to form a sulphuric-acid aerosol. Together with the smoke from the forest fires, this would have darkened the skies, so interfering with plant photosynthesis and disrupting the world's ecosystems, and also causing an initial global cooling. But this cooling would have been followed by a more long-lasting greenhouse effect due to the large quantities of CO_2 released by the heating of the carbonate rocks. This might have lasted for several tens of thousands of years, until the CO_2 could be absorbed by the oceans.

These deductions are supported by botanical studies that show a sudden reduction in the amount of flowering plant pollen at the K/T boundary in both North America and Japan. Above the soot-rich boundary clay in North America lies a layer that suggests the presence of rotting or burnt vegetation, with no pollen, and then a sudden, short-lived increase in the amount of fern spores [23]. This pattern of change is characteristic of the sequence of plant communities that in turn colonize an area after a forest fire. In the middle northern latitudes there was a particularly high rate of extinction in warmth-loving broad-leaved evergreen forest taxa, together with the survival and expansion of deciduous northern taxa — a pattern suggesting the extinction of taxa that were vulnerable to darkness or to cold, or to both. This abrupt climatic change was followed by a new stable climatic regime that continued well into the Paleocene.

Though, surprisingly, there is no trace in the marine plankton record of any temperature increase at the K/T boundary, this record does show that many types of plankton died out at this time, and marine productivity remained low for 0·5−1·0 million years [8]. Like the terrestrial plants] they would have been affected by the long darkness. So, both on land and in the oceans, many types of plant died out, unable to adapt to the sudden climatic change, and the effects of this would have risen through the food chain. As plant food became scarce, larger herbivores would have become extinct, followed by the larger carnivores that preyed upon them. The death of the plant plankton would similarly have been followed by that of the animal plankton that fed upon them. On land, only the smaller, cold-blooded reptiles and the little mammals that did not make major demands upon their environment would have found it easy to survive the K/T catastrophe.

The rise of the flowering plants

Angiosperms are first known in the Early Cretaceous, 120 million years ago. Many modern angiosperm families are recognizable in the Northern Hemisphere by 95 million years ago, at the Early/Late Cretaceous boundary, so differentiation of the flowering plants seems to have been fairly rapid. Though our knowledge of their early distribution is heavily dependent on records from the lowlands of the Northern Hemisphere, the pattern consistently suggests that they first diversified and became dominant in low latitudes, up to 20° from the Equator [24−26]. The first angiosperms may have been early successional herbs or shrubs in areas recently disturbed by erosion and deposition, only later diversifying

ecologically to occupy stream-side and aquatic habitats, the forest
understory and early successional thickets [27]. Even at the end of the
Cretaceous, though angiosperms formed 60–80 per cent of the low-
latitude floras, they comprised only 30–50 per cent of those in high
latitudes. The rise of the angiosperms was paralleled by a corresponding
reduction in the numbers and variety of mosses, club-mosses, horsetails,
ferns and cycads, but there was less change in the overall diversity of
conifers (Fig. 7.4).

Understanding the dispersal of the flowering plants presents fewer
problems than understanding that of mammals, for it commenced earlier,
before the break-up of the supercontinents had progressed very far. If
they first evolved and diversified in low, near-equatorial latitudes, the
tropical flowering plants would have been able to spread eastwards
through Africa and thence eventually into southern Eurasia. They would
also have been able to spread northwards into North America and from
there, once they had colonized the higher latitudes of the north, across
the still dry and warm Bering region between Alaska and Siberia, into
northern Eurasia. Even in the Southern Hemisphere, those angiosperms
that adapted to the high south-latitude cool-temperate region were able
to disperse through the South America–Antarctica–Australia land mass
before this became divided up (see p. 185) [28].

In the Middle Cretaceous, it is clear from pollen records that in
Gondwana, as in Laurasia, separate Northern and Southern floral
provinces had differentiated, so that broadly latitudinal floras now
existed in both Hemispheres (Fig. 7.5b) [29]. These comprised a humid
temperate Northern Laurasian flora dominated by the endemic conifer
family Pinaceae and other conifers, with some ferns; a rather similar
Southern Laurasian flora with more ferns and with conifers other than
the Pinaceae; a tropical Northern Gondwana flora with many cycads
and ephedras and few ferns; and a humid Southern Gondwana flora with
many podocarp conifers and many ferns. (Whether or not this last flora
extended into southern Africa or India, already isolated from the rest of
Gondwana by ocean, is unknown.)

This was the pattern of the floral provinces into which and through
which the flowering plants spread and diversified, apparently beginning
in the tropical regions of the Southern Laurasian and Northern Gondwana
provinces. In the Southern Hemisphere, a unique Southern Gondwana
angiosperm flora evolved, including mainly evergreen trees (such as the
southern beech tree, *Nothofagus*), and various shrubs and herbs, which
lived alongside the existing podocarp conifers and ferns. The total flora
was very like that of New Zealand today, and covered the southern
parts of South America plus Antarctica, Australia and New Zealand, but
it is not known from southern Africa, by then separated from the rest of
Gondwana by a wide stretch of ocean.

By the Late Cretaceous, the Northern Gondwana province contained a diversity of pollen of types now found in palms, but the floras of tropical South America and West Africa were becoming progressively more different as the two continents moved apart; they had become distinctly different by the Eocene. A more complicated pattern appeared in the Northern Hemisphere, where the latitudinal pattern of Northern Laurasian versus Southern Laurasian floras was affected by the longitudinal division of the land mass into Asiamerica and Euramerica (Fig. 7.5c). Pollen belonging to the group known as *Aquilapollenites* is found throughout the Asiamerican region, between the Obik Sea at the western end of Siberia and the Mid-Continental Seaway that bisected North America, but it is also found in the northern part of Euramerica. Pollen that belongs to the group known as *Normapolles* (which may be related to the hamamelid group of flowering plants, which includes the witch hazel) has a more restricted distribution, being found only in the warmer south-eastern parts of North America and the scattered islands that then made up southern Europe. This flora contains a greater abundance of flowering plants than the *Aquilapollenites* flora.

Late Cretaceous and Cenozoic climatic changes

There is much evidence that the climate of the world in the Mesozoic was very different from that of today. Oxygen-isotope measurements of the composition of the fossil shells of marine Cretaceous plankton (see p. 255) show that the intermediate to deep waters of these oceans were 15°C warmer than those of today. On land the presence of plants, dinosaurs and early mammals in high latitudes, and their spread through high-latitude routes such as the Bering region and Antarctica, support these observations. But from the Late Cretaceous onwards the earth's climate changed in two distinguishable, but linked, ways: it became cooler, changing from a generally warm world to one with polar ice sheets, and it became more seasonal. Several different factors can be identified that have influenced these changes, but we still do not fully understand precisely how these factors combined to produce changes of the magnitude and pattern that our analyses reveal [8].

An important factor has been the changing patterns of land and water, in which both the comparatively shallow epicontinental seas and the deep oceans had effects on the climate. Water absorbs more heat than does the land, and releases it more gradually, so slowing the intensity of any developing seasonal cycles. When, in the Early Cenozoic, the seaways that had subdivided North America and Asia withdrew, the climates of the central parts of those continents must have become less equable and more seasonal, with hotter, drier summers and colder, wetter winters. The deeper oceans not only affect the climates of adjacent land areas directly, but also affect climatic patterns more generally, for the ocean currents redistribute heated or cooled water masses, transporting them to other parts of the globe.

Both these oceanic influences are affected by the sizes and positions of the continents. For example, the separation of Australia from Antarctica 40 million years ago allowed the development of a great Antarctic circumpolar current. This current's persistently cold water has had a great influence on the origin and development of Antarctic ice and on the increasing aridity of Australia (see p. 217), because it has prevented the warmer waters of the South Pacific from penetrating further southwards. In contrast, although cold, deep Arctic water has been able to enter the North Atlantic since the opening of the Norwegian Sea between Greenland and Scandinavia in the Late Eocene, the lands surrounding the North Atlantic have been affected by the warm surface waters of the Gulf Stream, channelled northwards by the east coast of North America. The Kuroshio current produces a similar effect in the North Pacific, warming Japan.

The changing distribution of land and ocean as Africa and India moved northwards towards and into Eurasia, shrinking and finally obliterating the oceans that had previously lain between them, is also likely to have affected the amount of solar radiation that the earth absorbed. This radiation is more effectively absorbed at low latitudes (where it is more vertical to the surface) than at high latitudes, and is more efficiently absorbed by water than by land. So the fact that, over the last 60 million years, the area of land in low latitudes has increased very greatly, especially in the Northern Hemisphere, is likely to have had a perceptible influence on temperature. Combined with the greater seasonality of the northern continents after the epicontinental seas withdrew, this may have been an important factor in causing the declining temperatures that culminated in the appearance and spread of ice sheets at high latitudes and altitudes in the Late Cenozoic.

Another important factor was the rise of more mountain ranges during the Cenozoic. Mountains that rise 2–3 km above sea level not only themselves gather a covering of ice and snow, but also act as barriers to wind flow. They therefore cause the appearance of rain shadows in their lee, and also act as a dam to the movement of air that has been cooled in mid-latitude interiors in the winter. In North America, though the Rockies started to rise in the Triassic Period, the rate of this increased in the Late Eocene and they did not reach their present height until the Miocene or Pliocene; further west, the Sierra Nevada, Cascade and Coast ranges only started to rise over the last five million years. All these mountains will have increased the seasonality of the climate of North America, and the comparatively recent (Quaternary) rise of the Himalayas to their present elevation will similarly have held back the northern air mass that has cooled during the Asian winter. The effects of the rise of the Andes in South America during the Cenozoic were less severe, because the continent as a whole is narrower and is mainly at a lower latitude.

It has recently been suggested that the final onset of the cooling that led to the Ice Ages was caused by the major uplift of the mountains and high plateaux of western North America, and of the Himalayas and Tibetan plateau since the early Pliocene [30]. Climatic modelling experiments show that, by diverting the westerly winds of the Northern Hemisphere into a more northerly track, the appearance of these topographic features would have led to precisely the climatic effects that are observed to have taken place. These include winter cooling of much of the northern part of the Northern Hemisphere, summer drying of the North American west coast and of the Eurasian interior, and winter drying of the North American northern plains and of the interior of Asia.

The long-term cooling of the earth (Fig. 7.6a) led eventually to the appearance of polar ice sheets, commencing 40–10 million years ago in the Antarctic, surrounded by the cold Circumpolar current, and from 3 million years ago in the Northern Hemisphere. Ice reflects the sun's rays back into space, so the appearance of these ice caps will in itself have contributed to further climatic deterioration. The removal of this amount of water from the sea led to falling sea levels (Fig. 7.6b). This, together with the cooling of the ocean itself, with consequently decreased evaporation to provide rain, also led to increased aridity of the continents, especially in mid-latitudes, which can be detected from the amount of wind-blown dust deposited in deep-sea sediments (Fig. 7.6c).

Despite our recognition of all the above factors, and our ability nowadays to produce complex computer-generated models of the effects of alterations in their magnitude or location, it is still not possible to construct models that accurately match the high temperatures of the Cretaceous and Early Cenozoic, especially in the continental interiors. This at present can only be explained by making additional *ad hoc* assumptions of, for example, higher solar output of radiation or higher levels of atmospheric CO_2 in these earlier periods, although no convincing evidence of these phenomena has yet been identified.

Late Cretaceous and Early Cenozoic floral changes

Although, as already noted, there is a variety of different sources of evidence as to the world's past climatic regimes, the fossilized leaves of flowering plants provide an additional source after they became common and diverse, in the Middle Cretaceous. In areas of high mean annual temperature and rainfall the leaves have 'entire' margins, not

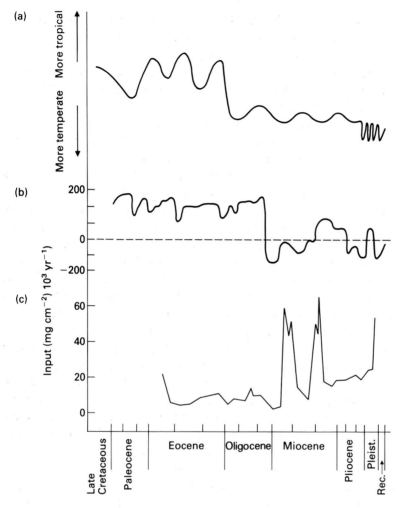

Fig. 7.6 The climatic record of the Late Cretaceous and Cenozoic. (a) Temperature, as suggested by flora from middle latitudes, after Wolfe [31]. (b) Generalized history of sea-level changes, relative to present day (0). (c) Amount of wind-borne dust in deep-sea sediments. (b) and (c) after Tallis [35].

subdivided into lobes or teeth; they are large and leathery, and are often heart-shaped, with tapering pointed tips, and a joint at the base of the leaf. These features are less common in floras from areas of low mean annual temperature and rainfall. Further information can be gained from the climate preferred by the living representatives of the families of plant found in each flora — though this line of deduction must be treated with caution, in case these climatic preferences have changed over time, or as a consequence of new competitive interaction. (For example, it is quite possible that the restriction we see today of ferns to moister situations may be because of competition from flowering plants, rather than from any inherent inability to colonize drier environments.)

A further danger is that the fossil data are often biased towards near-coastal areas of deposition, whose climates are usually milder and less seasonal than those of more inland areas.

The cooling of the world's climate began in the Middle Cretaceous, and this is clearly shown in a series of floras, ranging over 30 million years of that time, from about 70°N in Alaska [32]. The earliest of these floras contains the remains of a forest dominated by ferns and by gymnosperms such as cycads, ginkgos and conifers. The nearest living relatives of this flora are found in forests at moderate heights in warm-temperate areas at about 25–30°N — for example, in South-East Asia. By the time of the last of these Middle Cretaceous Alaskan floras, the flora had changed in two ways. Firstly, the angiosperms had by this time diversified to such an extent that they dominated the flora. Secondly, this forest contains the remains of forest similar to that found today at a latitude of 35–40°N in the region of North China and Korea — much further north than the living relatives of the earlier flora. The differences between these successive fossil floras therefore suggest that the climate of northern Alaska was already becoming cooler in the Middle Cretaceous. Other floras from the same area show that, by the end of the Cretaceous, the mean temperature had dropped by c. 5°C, and that the diversity of the flowering plants had dropped very greatly [33]. However, the climate of the main part of the North American continent appears to have changed less during this time [34]. At the end of the Cretaceous, there seems to have been a brief period of lower temperatures, perhaps associated with the extinction event at that time, followed by a longer-lasting increase in rainfall in North America.

All the known Early Cenozoic floras were dominated by woody plants, so forests clearly extended close to the poles in this warm, ice-free world [35]. It is therefore not surprising to find that taxa that are today restricted to particular latitudes were, in the Eocene, found 40–50° further from the Equator, or existed over a much wider band of latitudes. Tropical or subtropical floras are found in northern North America and Eurasia, extending as far north as floras of this age are known, to latitudes of 50°N in Europe and 65°N in North America, and as far south as Borneo. Most of the Early Cenozoic flowering plants were therefore what are today called megatherms and mesotherms (preferring respectively mean annual temperatures of above 20°C and between 20°C and 13°C); there were few microtherms (those preferring mean annual temperatures of below 13°C).

The climate improved again 50 million years ago, in the Early Eocene. This appears to have been caused by a sudden change in the temperature of the deep sea, for the American oceanographers James Kennett and Lowell Stott [36] have found that, around Antarctica, this increased at that time from 10°C to 18°C in less than two thousand years. They suggest that this change may have been due to a reversal of the normal pattern of oceanic circulation, in which cold water sinks at high latitudes and returns to the surface at low latitudes. But the possible cause of such a reversal is unclear, and it is also difficult to explain the fact that such a fundamental change in oceanic circulation lasted for only 30 000 years (but see p. 279). An example of the Early Eocene warming is southern England, then at a latitude of about 45°N. The fossil seeds and fruits of 350 species of plant, belonging to over 150 genera, are preserved in the London Clay deposits near the mouth of the River Thames [37]. The flora includes magnolias, vines, dogwood, laurel, bay and cinnamon, as well as the palm-trees *Nipa* and *Sabal* and the conifer *Sequoia*. Some of these plants are now found in the tropics, especially in Malaysia and Indonesia, while others live in temperate conditions like those of east-central China today. The closest analogue today would be a subtropical rain forest but, like that of many other Early Cenozoic floras, the composition of the London Clay flora is not identical to that of any single modern flora. The accompanying fauna, which includes crocodiles and turtles, is similar to that found today in the tropics.

Polewards of this Northern Hemisphere subtropical flora lay a rather narrow band of evergreen forest. An example of these high-latitude warm to warm-temperate floras is that found in Eocene deposits in Ellesmere Island in the Canadian Arctic, at a palaeolatitude of over 80°N, accompanied by a diverse fauna of mammals, lizards, snakes, turtles and salamanders [38,39]. The polar region itself was surrounded by broad-leaved deciduous angiosperm forest, together with character-istically northern needle-leaved conifers such as pine, larch, fir and spruce.

In the Southern Hemisphere, the floras of South America, Antarctica and Australia still contained, as they do today, descendants of the old Late Cretaceous Southern Gondwana angiosperm flora, with such families as the proteas, myrtles, *Nothofagus* and the 'southern conifers' such as *Araucaria*, *Podocarpus* and *Dacrydium*. These forests covered at least the periphery of Antarctica, but the inland flora of that continent is unknown.

The Early Eocene climatic improvement was, however, comparatively temporary, for studies of the leaf characteristics of the Northern Hemisphere floras of the Cenozoic show that there was a rapid climatic cooling during the Late Eocene and Early Oligocene (Fig. 7.6a). This involved both a decrease in the mean annual temperature, and an in-crease in the mean annual range (i.e. a decrease in equability). For example, within one or two million years the mean annual temperature of the Pacific North-West of North America dropped from about 27°C to about 12°C, while the mean annual temperature range increased from about 5°C to nearly 25°C. In this short space of time the broad-leaved evergreen forests of middle to high latitudes were replaced by temperate broad-leaved deciduous forests. It is not surprising to find that the floral diversity became much lower at this time, for many families must have become extinct, in these areas at least.

One result of the increasing seasonality of the climate was that the old closed evergreen forests became thinner and deciduous, providing an evolutionary opportunity for the appearance of new types of herbivorous, non-woody flowering plant. Another opportunity arose with the spread of arid environments. In the Late Eocene, the warm seas had easily evaporated to provide moist, rain-bearing winds, and there were not yet regional zones of arid and semi-arid climate in the subtropics, where today the high-pressure areas of a descending dry air produce deserts (see p. 80), although there were deserts in the continental interiors. But, after the Late Eocene/Early Oligocene cooling, the colder seas provided less rain, and there was a great expansion of arid-land vegetation [40]. Many flowering plant families that are common in such environments, such as the Brassicaceae and Plantaginaceae, appeared in the Late Miocene, as well as an increasing variety of the Asteraceae and Agavaceae and of the grasses. This vegetational change in turn provided an evolutionary advantage for the ruminant types of hoofed mammal, which can make maximum use of a limited amount of vegetation because they can digest even the cell walls of their plant food [41]. So both the plant and the animal world made evolutionary innovations in response to the world's cooling climates.

Towards today

After the Eocene, all parts of the world continued to be affected by climatic changes. For example, in the Late Miocene, 5−7 million years ago, the Antarctic ice sheet increased its size by up to 50 per cent. This caused the Antarctic circumpolar current to penetrate further north-wards, cooling the more northern regions, even as far as the Northern Hemisphere, and perhaps being responsible for drying of the climate in western North America. A couple of million years later, in the Early Pliocene, there was a considerable reduction in the size of the ice sheet, allowing these areas to become warmer once more. But the continents and their living inhabitants were becoming more and more separated from one another, and the impacts of climatic changes on each continent were also different, depending on its size and latitude. It is therefore more profitable to turn now to examine the major patterns of animal and plant realms on the planet today, and to follow the histories of each of these regions from the Early Cenozoic onwards.

References

1 Wegener, A. (1929) *The Origin of Continents and Oceans* (1966 English translation of 4th edition). Methuen, London.

2 Cox, C.B. (1974) Vertebrate palaeodistributional patterns and continental drift. *Journal of Biogeography* **1**,75–94.

3 Scotese, C.R. & McKerrow, W.S. (1990) Revised world maps and introduction. In *Palaeozoic Palaeogeography and Biogeography* (eds W.S. McKerrow & C.R. Scotese). Geological Society Memoir No. 12, pp. 1–21.

4 Stanley, S. (1986) *Earth and Life through Time.* W.H. Freeman, New York.

5 Edwards, D. (1990) Constraints on Silurian and Early Devonian phytogeographic analysis based on megafossils. In *Palaeozoic Palaeogeography and Biogeography* (eds W.S. McKerrow & C.R. Scotese). Geological Society Memoir No. 12, pp. 233–242.

6 Young, G.C. (1990) Devonian vertebrate distribution patterns and cladistic analysis of palaeogeographic hypotheses. In *Palaeozoic Palaeogeography and Biogeography* (eds W.S. McKerrow & C.R. Scotese). Geological Society Memoir No. 12, pp. 243–255.

7 Milner, A.R. (1990) The radiations of temnospondyl amphibians. In *Major Evolutionary Radiations* (eds P.D. Taylor & G.P. Larwood). Systematics Association Special Volume No. 42, pp. 321–349.

8 Crowley, T.J. & North, G.R. (1991) *Paleoclimatology. Oxford Monographs in Geology and Geophysics,* **8**,1–339. Oxford University Press, New York.

9 Witzke, B.J. (1990) Palaeoclimatic constraints for Palaeozoic palacolatitudes of Laurentia and Euramerica. In *Palaeozoic Palaeogeography and Biogeography* (eds W.S. McKerrow & C.R. Scotese). Geological Society Memoir No. 12, pp. 57–73.

10 Ziegler, A.M. (1990) Phytogeographic patterns and continental configurations during the Permian Period. In *Palaeozoic Palaeogeography and Biogeography* (eds W.S. McKerrow & C.R. Scotese). Geological Society Memoir No. 12, pp. 367–379.

11 Crowley, T.J., Hyde, W.T. & Short, D.A. (1989) Seasonal cycle variations on the supercontinent of Pangaea. *Geology* **17**,457–460.

12 Batten, D.J. (1984) Palynology, climate and the development of Late Cretaceous floral provinces in the Northern Hemisphere: a review. In *Fossils and Climate* (ed. P. Brenchley), *Geological Journal,* Special Issue No. 11, pp. 127–164. John Wiley, London.

13 Chaloner, W.G. & Lacey, W.S. (1973) The distribution of Late Palaeozoic floras. *Special Papers in Palaeontology* **12**,271–289.

14 Cox, C.B. (1973) Triassic tetrapods. In *Atlas of Palaeobiogeography* (ed. A. Hallam). Elsevier, Amsterdam, pp. 213–223.

15 Cox, C.B. (1967) Changes in terrestrial vertebrate faunas during the Mesozoic. In *The Fossil Record* (ed. W.B. Harland). Geological Society, London, pp. 71–89.

16 Veevers, J.J., Powell, C. McA. & Roots, S.R. (1991) Review of seafloor spreading around Australia. I. Synthesis of the patterns of spreading. *Australian Journal of Earth Sciences* **38**,373–389.

17 Woodburne, M.O. & Zinsmeister, W.J. (1984) The first land mammal from Antarctica and its biogeographic implications. *Journal of Paleontology* **58**, 913–948.

18 Pascual, R. *et al.* (1992) First discovery of monotremes in South America. *Nature* **356**,704–706.

19 Godthelp, H. *et al.* (1992) Earliest known Australian Tertiary mammal fauna. *Nature* **356**,514–516.

20 Alvarez, L.W. *et al.* (1980) Extraterrestrial cause for the Cretaceous–Tertiary

extinctions. *Science* **208**,1095–1108.

21 Sigurdsson, H. *et al.* (1991) Glass from the Cretaceous/Tertiary boundary in Haiti. *Nature* **349**,482–487.

22 McKinnon, W.B. (1992) Killer acid at the K/T boundary. *Nature* **357**,15–16.

23 Spicer, R.A. (1989) Plants at the Cretaceous–Tertiary boundary. *Philosophical Transactions of the Royal Society of London* **B325**,291–305.

24 Crane, P.R. & Lidgard, S. (1989) Angiosperm diversification and paleolatitudinal gradients in Cretaceous floristic diversity. *Science* **246**,675–678.

25 Crane, P.R. & Lidgard, S. (1990) Angiosperm radiation and patterns of Cretaceous palynological diversity. In *Major Evolutionary Radiations* (eds P.D. Taylor & G.P. Larwood). Systematics Association Special Volume No. 42, pp. 377–407.

26 Lidgard, S. & Crane, P.R. (1990) Angiosperm diversification and Cretaceous floristic trends: a comparison of palynofloras and leaf macrofloras. *Paleobiology* **16**,77–93.

27 Crane, P.R. (1987) Vegetational consequences of the angiosperm diversification. In *The Origins of Angiosperms and their Biological Consequences* (eds E.M. Friis, W.G. Challoner & P.R. Crane). University Press, Cambridge, pp. 107–144.

28 Drinnan, A.N. & Crane, P.R. (1990) Cretaceous paleobotany and its bearing on the biogeography of austral angiosperms. In *Antarctic Paleobiology* (eds T.N. Taylor & E.L. Taylor). Springer Verlag, New York, pp. 192–219.

29 Batten, D.J. (1984) Palynology, climate and the development of Late Cretaceous floral provinces in the Northern Hemisphere: a review. In *Fossils and Climate* (ed. P. Brenchley). *Geological Journal*, Special Issue **11**,127–164. Wiley, London.

30 Ruddiman, W.F. & Kutzbach, J.E. (1989) Forcing of Late Cenozoic Northern Hemisphere climate by plateau uplift in southern Asia and the American West. *Journal of Geophysical Research* **94D**,18405–18427.

31 Wolfe, J.A. (1978) A paleobotanical interpretation of Tertiary climates in the Northern Hemisphere. *American Scientist* **66**,694–703.

32 Smiley, C.J. (1966) Cretaceous floras from Kuk River area, Alaska; stratigraphic and climatic interpretations. *Bulletin of the Geological Society of America* **77**,1–14.

33 Parrish, J.T. & Spicer, R.A. (1988) Late Cretaceous vegetation: a near-polar temperature curve. *Geology* **16**,22–25.

34 Wolfe, J.A. & Upchurch, G.R. (1987) North American nonmarine climates and vegetation during the Late Cretaceous. *Palaeogeography, Palaeoclimatology and Palaeoecology* **61**,33–77.

35 Tallis, J.H. (1991) *Plant Community History. Long-Term Changes in Plant Distribution and Diversity.* Chapman & Hall, London.

36 Kennett, J.P. & Stott, L.D. (1991) Abrupt deep-sea warming, palaeoceanographic changes and benthic extinctions at the end of the Palaeocene. *Nature* **353**, 225–229.

37 Collinson, M.E. (1983) *Fossil Plants of the London Clay.* Palaeontological Association, London.

38 McKenna, M.C. (1980) Eocene paleolatitude, climate, and mammals of Ellesmere Island. *Palaeogeography, Palaeoclimatology and Palaeoecology* **30**,349–362.

39 Estes, R. & Hutchinson, J.H. (1980) Eocene lower vertebrates from Ellesmere Island, Canadian Arctic Archipelago. *Palaeogeography, Palaeoclimatology and Palaeoecology* **30**,325–347.

40 Singh, G. (1988) History of aridland vegetation and climate: a global perspective. *Biological Reviews* **63**,159–195.

41 Janis, C.M. (1989) A climatic explanation for patterns of evolutionary diversity in ungulate mammals. *Palaeontology* **32**,463–481.

8 Patterns of life today

The commonly accepted patterns of major division of the world into faunal and floral regions are shown in Figs 8.1 and 8.3. Naturally enough, these divisions have been based upon the distribution of dominant, easily visible groups — birds and, later, the mammals, and the flowering plants. The biogeographic patterns of older, less dominant groups are complicated by the frequent presence of relict distributions (see p. 38). They are therefore less easy to interpret, and contribute less to any general understanding of world biogeography. Nevertheless, valuable information has been published on lower land plants [1], invertebrates in general [2], insects [3], fish and lower land vertebrates [4], anurans [5], and birds [6].

The nature of the biota of mammals and of flowering plants that developed on each continent was the result of the interaction between the early history of these two groups, the gradual fragmentation of the continents (especially in the Southern Hemisphere) and the climatic changes that took place during the Cenozoic. These early, but crucial, aspects of historical biogeography are therefore considered next, before their detailed results in each region are examined.

Mammals — the final patterns

Because both South America and Africa were largely isolated from other continents in the Early Cenozoic, each developed a characteristic mammalian fauna. This is reflected in the long-established zoogeographic map of the world (Fig. 8.1). India and South-East Asia at one time had a fauna similar to that of Africa, but their different climatic histories, and later separation by desert and mountains, have led to a divergence of their mammalian faunas. As a result, a separate Oriental zoogeographic region is recognized. The division between this region and the Australian region, with its unique marsupial fauna, lies in the East Indian island area that both faunas are still colonizing. The two land masses of the Northern Hemisphere have mammalian faunas that differ somewhat from one another, but both are similar in having been greatly impoverished by the climatic change of the Pleistocene Ice Ages.

The result of all the processes of evolutionary change and of dispersal at different times and by different routes has been that several patterns of distribution can be seen among the orders of mammal. The pattern in the Late Cenozoic Miocene–Pliocene Epochs is shown in Table 8.1. The final pattern found today is slightly different from this, because of

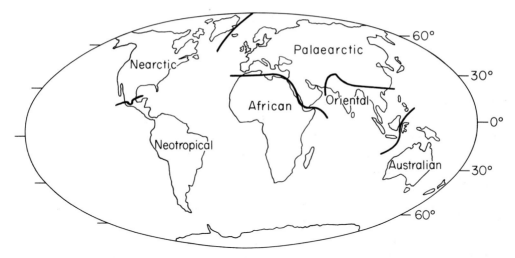

Fig. 8.1 Zoogeographical regions of the world today, based on the distribution of mammals.

the extinction of elephants in the Palaearctic and Nearctic during the Pleistocene, and because of the dispersal of edentates and marsupials to the Nearctic via the Panama land bridge. The final total of orders for each region takes account of these changes. The last line of Table 8.1 also shows the total number of terrestrial families of mammal in each

Table 8.1 The distribution of the orders of terrestrial mammals during the late Cenozoic (Miocene–Pliocene). The final total of orders also takes account of Quaternary extinctions and dispersals (see text).

	Africa	Orient	Pal.	Nearc.	Neotrop.	Austr.
Rodents	×	×	×	×	×	×
Insectivores, carnivores, lagomorphs	×	×	×	×	×	
Perissodactyls, artiodactyls, elephants	×	×	×	×	×	
Primates	×	×	×		×	
Pangolins	×	×				
Conies, elephant-shrews, aardvarks	×					
Edentates					×	
Marsupials					×	×
Monotremes						×
Total number of orders today	12	9	7	8	9	3
Total number of terrestrial families today	44	31	29	23	32	11

region; these figures therefore exclude whales, sirenians, pinnipedes (seals, etc.), bats, and also humans and their domestic passengers (such as the dingo and rabbit in Australia).

Within each order of mammal, the individual families show considerable variations in their success at dispersal. A few have been extremely successful. Nine families have dispersed to all except the Australian region: soricids (shrews), sciurids (squirrels, chipmunks, marmots), cricetids (hamsters, lemmings, voles, field mice), leporids (hares and rabbits), cervids (deer), ursids (bears), canids (dogs), felids (cats), and mustelids (weasels, badgers, skunks, etc.). In addition, the bovids (cattle, sheep, impala, eland, etc.) have dispersed to all except the Neotropical and Australian regions, and the murids (typical rats and mice) have dispersed to all except the Neotropical and Nearctic regions. This group of 11 families can conveniently be called 'the wanderers'. Their inclusion in any analysis of the patterns of distribution of the living families of terrestrial mammal tends to blur the underlying patterns of relationship of these zoogeographic regions. These 'wanderers' have therefore been excluded from Fig. 8.2, which shows the distribution of the remaining 79 families. The Oriental region is shown twice, so that the single family shared with the Neotropical region (the relict distribution of camelids) can be included.

As can be seen from Fig. 8.2, the majority of all the families of terrestrial mammal (51 out of 90, i.e. 57 per cent) are endemic to one region or another. The degree of endemicity of the mammals in each of the different regions is calculated in Table 8.2. (It is worth noting that the rodents, the most successful of all the orders of mammal,

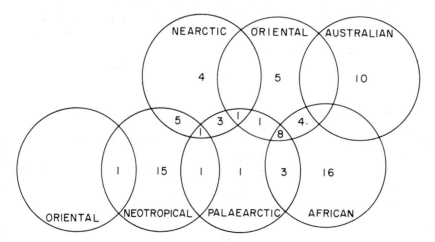

Fig. 8.2 Venn diagram showing the inter-relationships of the families of terrestrial mammal of the six zoogeographic regions today, excluding the 11 'wandering' families.

contribute 19 of these endemic families: two Nearctic, one Palaearctic, ten Neotropical and six African.) It is clear from Fig. 8.2 and Table 8.2 that the degree of distinctiveness of the six zoogeographic regions of today, if judged by the endemicity of their mammals, varies greatly. These figures are the resultant of three main factors: isolation, climate, and ecological diversity.

The results of the long isolation of the Australian and Neotropical regions are obvious. The other four regions were all interconnected during the Mid-Cenozoic. The Nearctic region was connected to Eurasia via the high-latitude Bering region, and many mammal groups were able to disperse across this region during those warmer times. Nevertheless, these groups did not include the tropical and subtropical groups that then ranged throughout Africa and southern Eurasia — including parts of Eurasia well north of the present limits of the Oriental region, which was therefore not recognizable as a separate zoogeographic region at that time. The Pleistocene glaciations of the Northern Hemisphere then decimated the mammal faunas of both the Nearctic and the Palaearctic. These two regions therefore contain few mammal families, and these families are mostly also still found in the adjoining southern regions, so that the Nearctic and Palaearctic contain few endemic mammal families. Tropical and subtropical Old World mammal families are therefore now found only in the Oriental and African regions. Though, without the three families found only in Madagascar, the degree of endemicity of the mammals of the African region would be slightly lower $(13 \times 100 \div 41 = 32$ per cent), it is still significantly higher than that of the Oriental region. It is interesting that this difference is found also in birds: 13 families are endemic to the African region, but only one is endemic to the Oriental region. These differences are probably the result of the greater diversity of environments in Africa, due to the Late Cenozoic spread of woodland in Africa (see p. 227), as well as to the greater size of the African continent.

It is also interesting to analyse the extent to which the different

Table 8.2 The degree of endemicity of the families of terrestrial mammal: number of endemic families × 100 ÷ total number of families.

Region	Endemicity
Australian	$10 \times 100 \div 11 = 91\%$
Neotropical	$15 \times 100 \div 32 = 47\%$
African	$16 \times 100 \div 44 = 36\%$
Holarctic	$7 \times 100 \div 37 = 19\%$
Nearctic	$3 \times 100 \div 23 = 13\%$
Oriental	$4 \times 100 \div 31 = 13\%$
Palaearctic	$1 \times 100 \div 30 = 3\%$

ecological niches have been filled in the different zoogeographical regions (Tables 8.3, 8.4). This has been outlined by Eisenberg [7] and shows, for example, that 21 per cent of the mammal genera of Africa are browsers or grazers, compared with only 9 per cent in South America. This is related to the presence of extensive grasslands and savanna in East Africa (see p. 227), while the influence of the great South American rain forests is shown by the fact that 22 per cent of the mammal genera of northern South America are fruit-eating or omnivorous, compared with only 11 per cent in southern Africa. For the same reason, northern South America also has a high proportion of mammals (bats) that feed on insects either in the air or on leaves; the small proportion of mammals in South-East Asia that feed on insects living on leaves is surprising.

Comparison of data of this kind provokes interesting questions. In Australia, 20 per cent of the mammalian genera are arboreal, as in South-East Asia, despite the much smaller extent of forests in Australia today. Eisenberg suggests that this may be an inheritance from Australia's more heavily forested past (see below), during which many marsupials became arboreal. Similarly, he notes the comparatively small number of fruit-eaters/omnivores in Australia, and speculates as to whether this is because there are fewer fruit trees in Australia, or because parrot-type birds there have been more successful in that niche than have the mammals. It is also possible that the Australian fruit trees, which are less taxonomically diverse than those of South America and live in a more seasonal climate, may be more seasonal than those of South America, so that fruit is not a reliable, year-round source of food. This may also be the reason why there are also many more flying mammals (all bats) in northern South America (see p. 12).

A further viewpoint on mammalian faunas in the zoogeographic regions on a geological time scale is provided by Flannery's observation that Australia has very few large carnivores [8]. As he points out, there may be several reasons for this. The area of Australia is smaller compared to that of the other continents; much of it is unproductive desert or semi-desert, and the vegetation in general is of low productivity because of Australia's impoverished soils (see below); finally, Australia's annual rainfall is also extremely variable because it is highly affected by the 'El Niño' events in the Pacific Ocean. All this will lead to wide fluc-tuations in the populations of Australian herbivores. As a result, the carnivores that prey on them, and whose population sizes are always lower than those of their prey, will be even more vulnerable to extinction. Flannery notes that Australia has an unusually high number of reptilian predators, such as pythons and varanid lizards, and had giant members of these groups in the Pleistocene Period, as well as a large land crocodile. He suggests that the reptiles, being cold-blooded, may have been better

Table 8.3 Numbers of mammalian genera (excluding cetaceans) in each way of life in each zoogeographic region. After Eisenberg [7].

Region	Burrowing	Semi-burrowing	Aquatic	Semi-aquatic	Flying (bats)	Terrestrial	Terrestrial & arboreal	Arboreal	Total
South-East Asia	3	3	5	4	42	44	24	34	159
Northern South America	2	7	4	14	69	66	19	38	219
Southern Africa	8	1	3	8	39	123	11	21	214
North America	10	19	11	5	16	51	14	2	128
North Asia	5	10	8	7	10	76	12	8	136
Australia	1	1	6	3	19	38	9	19	96
Europe	2	4	3	5	10	40	8	5	77
Middle East	1	5	—	2	17	50	6	7	88
India	1	—	3	2	22	45	9	8	90
Southern South America	2	6	7	6	5	35	8	1	70
Madagascar	—	1	—	1	15	14	7	13	51

Table 8.4 Numbers of mammalian genera (excluding cetaceans) in each feeding category in each zoogeographic region. After Eisenberg [7].

Region	Fish, squid	Carni- vores	Nectar	Grain	Crustacea, molluscs	Ants, termites	Insects Aerial insects (bats)	Insects on foliage (bats)	Insects/ omni- vores	Fruit/ omni- vores	Fruit & grain	Fruit & plants	Browsers	Grazers	Plankton	Blood	
South-East Asia	9	10	2	—	—	1	22	2	22	27	31	14	13	4	—	—	159
Northern South America	12	9	9	1	—	8	27	9	14	48	41	19	14	6	—	3	219
South Africa	8	19	3	1	—	2	23	3	21	24	45	18	29	17	—	—	214
North America	10	8	2	—	2	—	11	3	16	13	24	15	16	8	—	—	128
North Asia	13	8	—	—	1	—	8	2	12	12	33	22	15	10	—	—	136
Australia	5	7	3	—	1	2	11	2	17	12	9	8	11	8	1	—	96
Europe	7	7	—	—	—	—	8	2	4	7	18	7	11	6	—	—	77
Middle East	2	9	—	—	—	—	13	3	8	9	21	10	8	5	—	—	88
India	5	6	—	—	—	1	15	1	8	15	16	6	8	8	—	—	90
Southern South America	6	4	—	—	—	1	5	—	9	11	7	14	6	6	1	1	70
Madagascar	1	3	1	—	—	—	12	1	14	6	4	5	4	—	—	—	51

able to survive periods of starvation and so to retain a higher, safer level of population.

The distribution of flowering plants today

The patterns of distribution of living angiosperms have been described by the Russian botanist Armen Takhtajan [9] (Fig. 8.3), while the distribution maps in the book edited by a British botanist, Vernon Heywood [10], provide an excellent basis for an evaluation of their biogeography.

It would be simple to explain their biogeography if there were few families of flowering plant, if the history of the origin and diversification of each were understood, and if they conformed to a few straightforward patterns. Unfortunately, none of these ideals is realized. Firstly, there are many more flowering plants than there are mammals — some 300 living families and 12 500 genera have been described, compared with only 100 families and 1000 genera of living mammal. Secondly, the fossil record is of little help in the interpretation of their biogeographic history, for the taxonomy of modern angiosperms is based largely upon the characters of the flower, which is hardly ever preserved. Furthermore, although leaves, wood, seeds, fruit and pollen grains are preserved, they are rarely found in such unequivocal association that the palaeobotanist can be sure which originally belonged to the same type of plant. Even when a particular family (e.g. the Magnoliaceae) has been identified in the Late Cretaceous or Early Cenozoic on the basis of characters of preserved pollen or leaves, it is dangerous to extend the concept and to assume that these early flowering plants were identical to the living

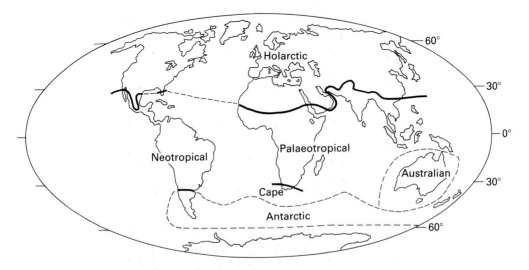

Fig. 8.3 Floral regions of the world today. After Takhtajan [9].

group in their appearance, habit, or ecological preferences. It is, therefore, not possible to identify the areas of origin and gradual evolutionary diversification and geographical spread of the different groups of flowering plant in the same way as those of the families of mammal.

Finally, the patterns of distribution of the flowering plant families are very diverse. This diversity may be partly because flowering plants are much better at dispersal across ocean barriers than are the mammals, since successful dispersal requires only a single air-borne seed instead of a breeding pair of mammals or a pregnant female. Certainly, most families of flowering plant are much more widely distributed than are most families of mammal. For example, almost everywhere in the world, four flowering plant families are among the six most numerous: the Compositae (daisies, sunflowers, etc.), Gramineae (grasses), Leguminosae (peas, clover, vetches, etc.) and Cyperaceae (sedges). Only 56 (18 per cent) of the families of flowering plant are endemic to one particular region (and most of these are relatively unimportant families restricted to only a few genera), while 51 (57 per cent) of the families of terrestrial mammal are similarly endemic. The more widespread nature of angiosperm distribution is also shown by the fact that, of the 302 families of flowering plant (excluding the four marine families), 86 are found world-wide, and 28 others are found in each region except the Boreal and Antarctic regions. It is therefore not surprising that it is difficult to characterize the different tropical regions.

Mammalian versus flowering plant geography — comparisons and contrasts

As might be expected, the systems of faunal and floral regions are very similar to one another (compare Figs 8.1 and 8.3). This is because the dispersals of both the mammals and the flowering plants began in the Middle or Late Cretaceous, and have therefore been limited by similar major significant differences between the two biogeographic systems. Since each is itself a generalization and simplification, it is particularly valuable to compare the two systems, for the recognition of differences and their explanation provide a useful test, and perhaps re-assessment, of these generalizations.

One obvious difference between the two systems lies in the southernmost Southern Hemisphere, where plant geographers recognize an Antarctic temperate floral region which excludes even the southern tip of Africa. This flora is a descendant of the Southern Gondwana flora already mentioned (p. 192), which covered the humid temperate parts of South America, Antarctica, Australia and New Zealand in the Late Cretaceous. Its forests consisted of mainly podocarp gymnosperms and mainly evergreen angiosperms, plus various herbaceous and shrubby angiosperms — a flora very like that of the New Zealand forests today. These Southern Hemisphere forests have hardly any deciduous species. Because it is mainly deciduous species that have entire-margined leaves (see p. 195), these forests also have a much lower proportion of such leaf-type species than those of the Northern Hemisphere [11]. Few of the angiosperms in this temperate flora were able to reach temperate southern Africa, because that part of the continent was already separated from the rest of Gondwana by widening oceans. Such angiosperms as *Nothofagus*, the southern beech tree, are therefore absent from Africa, although it is one of the most characteristic angiosperm trees of the southern temperate region, being found in southern South America, south-eastern Australia, New Caledonia, New Zealand and New Guinea. However, a few families (the Cunoniaceae, Dilleniaceae, Escalloniaceae, Gunneraceae, Petiveriaceae, Philesiaceae, Proteaceae and Restionaceae) from this flora are found also in Africa; these must have either been able to disperse to that area by land before the South Atlantic widened, or dispersed across that barrier.

Though much altered by subsequent climatic history, the flora of the Australian region is really also derived from the original Southern Gondwana flora (see p. 192). As far as mammals are concerned, only the presence of marsupials in both South America and Australia (and as fossils in Antarctica) reflects this ancient Gondwana link. This, then, is the reason for the very different biogeographic patterns of flowering plants and of mammals in the southernmost part of the Southern Hemisphere.

A second difference between the two biogeographic systems lies in the relationship between South-East Asia and Australasia (Fig. 8.4). The series of islands between the mainlands of Asia and Australia contains a transition between the Asian flowering plants and placental mammals, and the Australian flowering plants and marsupial mammals. The whole area is usually viewed by plant goegraphers as the Malaysian province of the Old World tropical floral region, which extends eastwards to include New Guinea. Zoologists, on the other hand, found that New Guinea contained marsupials, but very few placentals, and a predominantly Australian bird fauna. The zoogeographer Wallace had, in the nineteenth century, suggested a line of faunal demarcation, later called Wallace's Line, that separated the predominantly Asian bird fauna from the more eastern, predominantly Australian bird fauna. This Line, which runs close to the Asian continental shelf, is now normally recognized to be the boundary between the Oriental and the Australian zoogeographic regions. But, in reality, the area between the Asian and the Australian continental shelves (which biogeographers sometimes refer to as 'Wallacea') contains relatively few Asian or Australian mammals. As far as mammals

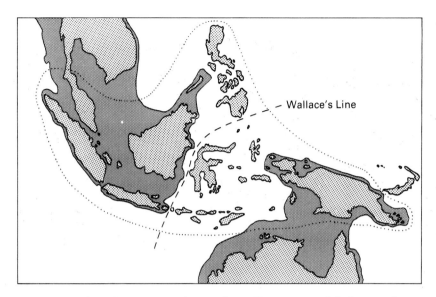

Fig. 8.4 Map of South-East Asia and Australasia. The continental shelves are shown lightly shaded. The Malaysian province is shown outlined by a dotted line. Wallace's Line indicates the boundary between the Oriental and Australian faunal regions.

are concerned, it would in many ways therefore be best omitted from the system of essentially continental vertebrate faunas on which the accepted faunal regions are based.

Such an area of exclusion from the zoogeographic regions might even be extended to include New Guinea, for its claim to Australian affinity depends on its bird fauna and on a very limited (21 genera) variety of Australian marsupials. Any system of zoogeographic regions that depended on other groups of animals would have to recognize that, for example, the insects of New Guinea are mainly of Asian origin.

A final major difference between the floral and the faunal regions is the recognition by botanists, but not by mammalian zoogeographers, of a separate Cape floral region confined to a small area of the southern tip of Africa. The historical plant geography of this little region is uncertain (see below).

In addition to these differences between the boundaries of the animal and plant systems of biogeography, there are also differences in the degree of closeness of relationship between the regions. As has already been noted, the relationships between the zoogeographic regions are seen more clearly after the deletion of the 11 families of 'wandering' mammal. Similarly, those of the flowering plant regions are also shown more clearly if the families with a world-wide distribution are first excluded. If these two patterns are now compared (Fig. 8.5), some very interesting differences can be seen.

Mammal families

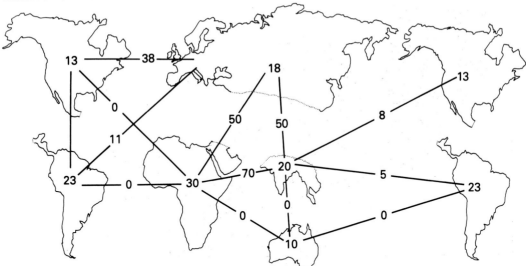

Flowering plant families

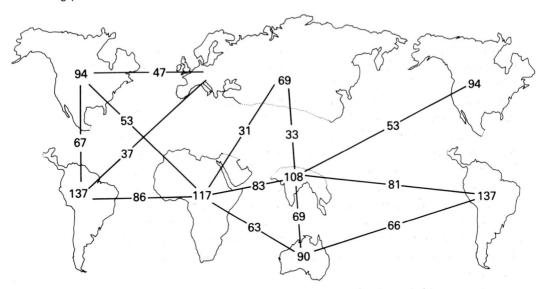

Fig. 8.5 A comparison of the faunal and floral similarities between the different regions. Figures within each region show the number of families found there. (Families found only in Madagascar have been omitted.) Figures linking the regions show the coefficients of biotic similarity (see p. 184) of the biota of these regions. Above, mammal families (excluding the 'wandering' families). Below, flowering plant families (excluding those with a world-wide distribution).

The first point is that there is a much greater similarity between the flowering plant floras of South America, Africa, the Oriental region and Australia than between the mammalian faunas of these regions; only in the case of the African/Oriental comparison are the plant and animal figures at a similar level (see below). Three factors seem to have caused these differences.

Firstly, the families of flowering plants evolved and dispersed earlier than the families of mammals. The fossil pollen record of 141 flowering plant families has been analysed by the Netherlands botanist Jan Muller [12], and this can be compared with the record of the 94 mammal families of which fossils are known. Though the opossums are the only mammal family (1 per cent of the total 94) known from the Cretaceous, 31 (22 per cent of the total 141) angiosperm families had already appeared by that time. In the Palaeocene, three families (3 per cent) of the mammals had appeared (hedgehogs, armadillos and horses), compared with 18 (13 per cent) of the flowering plants, and in the Eocene 15 (16 per cent) of the mammals and 33 (23 per cent) of the plants. Even allowing for the fact that the fossil record of the flowering plants is much less complete, so that many other families may have evolved later, it is still clear that the angiosperms had commenced their dispersal across the world much earlier than the mammals. They therefore had a much greater chance of colonizing the continent before they had drifted far apart. In contrast, mammals found it more difficult to colonize these continents at a later date, but those that did so then diverged into a number of unique, endemic groups that show little similarity to those in other continents — edentates, New World monkeys and caviomorph rodents in South America; elephants, elephant-shrews, conies and aardvarks in Africa; monotremes and marsupials in Australia.

The comparative youth of the mammalian families is at least partly because there has been a great deal of extinction and replacement during their history: in addition to the 100 living families of mammal, 140 other families evolved and became extinct during the Cenozoic. The fossil record of flowering plants is, as already explained, difficult to interpret. Because of the fragmentary and dissociated nature of that record, palaeobotanists have tended to place these remains into the ancestry of living families. Though this may have exaggerated the phenomenon, there does seem to have been comparatively little extinction of angiosperm families. These extinctions have therefore reduced the previous levels of faunal similarity, as can be seen most clearly by comparing the faunal similarities between North and South America before and after the Pleistocene extinctions (see pp. 219–223).

Finally, of course, it would be unwise to assume that all the floral similarities were merely the result of early dispersal across negligible barriers, rather than a later colonization across wider gaps. The extent of the spread of flowering plants across the Pacific (over 200 different immigrant flowering plants have reached the most isolated island group, Hawaii) shows clearly that they can cross even quite wide stretches of ocean, especially where intermediate island stepping-stones were available. So the fact that a number of angiosperm families, such as those mentioned on p. 211, dispersed to South America, Africa and Australia does not prove that all these families dispersed when the three areas were interconnected during the Cretaceous. Nevertheless, the number and variety of flowering plant families that are involved in the similarities between South America, Africa, the Oriental region and Australia suggest that it was common early inheritance, rather than later exchange by dispersal, that created the widespread common element in these floras.

As flowering plant families therefore started to disperse earlier than mammals, when the drifting continents were still closer together, as they have suffered less from extinction, and as they are better than mammals at dispersing, it is not surprising that the floras of the different continents show more similarities than do their mammalian faunas.

Though on the whole the floras of the different tropical continents are more similar than the faunas, there is one exception: the almost identical levels of similarity for the two groups when the African and Oriental regions are compared. The floral similarity here is not surprising, for it is at the same general level as the similarities between the other tropical regions — South America vs. the African region, and South America vs. the Oriental region. It is therefore the faunal similarity between the African and Oriental regions that is unusually high. The explanation for this lies, as we shall see below (p. 226), in the fact that these two regions were, probably as recently as the Miocene, connected by a broad stretch of tropical environment that covered what is now the Middle East. The similarities between their mammal faunas therefore reflect this Miocene relationship between Africa and India, rather than the divergence that began after their separation by deserts and by the Red Sea and Persian Gulf. The similarity between the levels of faunal and floral relationship does, however, highlight the dissimilarity between the normal practice of plant geographers, who recognize them as a single Palaeotropical floral region, and the zoogeographers, who separate them as separate African and Oriental regions.

The greater similarity, noted above, between the floras of the tropical regions also explains another difference. This is the fact that there is more similarity between the floras of North America and Eurasia, which are linked in a single Boreal floral region, than there is between the mammal faunas of these two regions. As we shall see below, both regions lost nearly all of their subtropical biota during the Ice Ages. After the last Ice Age had ended, both animals and plants started to spread northwards to re-colonize the newly warmed lands of the Northern

Hemisphere. However, there was already much more similarity between the plants of South America and Africa than between their mammals. Their dispersal into respectively North America and Europe therefore produced a corresponding similarity between the floras of these two regions. No such similarity resulted from the northward spread of the very different mammals of South America and of Africa.

Finally, the high similarity between the floras of South America and Africa is probably because a large number of South American tropical families have spread into the almost subtropical south-eastern region of North America.

The Cenozoic historical biogeography of each of the main regions, plus the large islands of New Zealand and Madagascar, will now be considered in turn.

Australia

The characteristics and biogeographic affinities of the biota of Australia are the most unusual and interesting of any in the world, and their explanation necessitates a very rewarding understanding of the interplay between continental movement, climatic change and biotic dispersal [13–15].

When flowering plants and mammals first evolved, radiated and dispersed, in the Cretaceous, Antarctica was still joined to both South America and Australia (Fig. 7.5). Warm climates extended to near the ice-free poles, so that the seas surrounding Antarctica were warm, and forests clothed at least its coastal regions as late as the Eocene. As already noted above, the Southern Gondwana flora that evolved in this whole region was adapted to a humid environment and included many podocarp conifers and also ferns. The flowering plants of this flora probably included the 14 families that are still mainly confined to the fragments of Gondwana (e.g. Proteaceae, Epacridaceae, Restionaceae, Monimiaceae) [16], and also such genera as the southern beech tree, *Nothofagus*. In Australia, most of these flowering plants are found today in rain forests — either in temperate cool rain forests or in warm, humid, seasonal rain forests. This suggests that this was also the original environment of the early Southern Gondwana flora of Australia.

The most unusual and characteristic flora of Australia is that known as the *sclerophyll* flora. The plants of this flora grow slowly, readily cease growth altogether, and have a small total leaf area that is composed of small, broad, evergreen leathery leaves. This flora is adapted to the unusual soils of Australia, which are highly weathered and low in nutrient minerals: those of the semi-arid zone have about half the levels of nitrates and phosphates of equivalent soils elsewhere. This is because the mountains of Australia are mainly old (c. 200 million years) and even these are only along the east coast. As a result there has been little erosion to add new sediments and minerals to the vast flat expanses of the continent. The plants of the sclerophyll flora seem to have evolved from the rain-forest flora, for all the larger families with sclerophyll types are also found in the rain forests, and about 45 per cent of the sclerophyll genera are endemic to Australia. The most spectacularly successful sclerophyll forms are the gum tree genus *Eucalyptus* (Myrtaceae), which includes about 500 species, and the family Proteaceae.

Though both the rain forest and the sclerophyll floras once covered much greater areas of Australia, they are today found only in isolated, scattered areas. This is because of Australia's history of continental drift (Fig. 8.6). After separating from Antarctica c. 40 million years ago, Australia moved northwards. At first the two continents were part of a single weather system, so that heat was transferred southwards from Australia to Antarctica. This ended in the Early Oligocene, c. 34 million years ago, when the two continents had separated sufficiently for a deep-water circumpolar current of cooler water and westerly winds to become established. This was the beginning of ice-formation in Antarctica. There was also less evaporation from the now-cooler seas around Australia, so that there was less rainfall on the continent and a consequent increase in its arid, desert areas. This process continued into the Miocene, by which time the aridity of central Australia led to the appearance of open grass-lands and savanna containing the varied genus *Acacia* (Leguminosae). Australia's continued northward drift took it, in Late Miocene times, into the 30°s high-pressure zone of low rainfall (see p. 82), further increasing its aridity. All these factors have therefore combined to make Australia the driest of all continents, two-thirds of it having an annual rainfall of less than 500 mm, and one-third having less than 250 mm. Here, then, is the reason for the restriction of its rain forest and sclero-phyll floras to isolated scattered areas.

This was also the climatic and vegetational history to which the mammals of Australia had to adapt. The early history of even the marsupials of Australia is unknown, for their earliest fossils are of mid-Miocene age, when the grasslands were starting to appear. That early marsupial fauna contained more browsers than the modern fauna, which is mainly composed of grazers feeding on the great expanses of grassland. The most diverse of these grazers are the kangaroos and wallabies but, in the isolation of Australia, marsupials have radiated into a great variety of forms, occupying the niches that placentals have filled every-where else. Marsupial equivalents of rats, mice, squirrels, jerboas, moles, badgers, ant-eaters, rabbits, cats, wolves and bears all look very like their placental counterparts.

The sclerophyll vegetation of Australia is low in nutrients and high in toxic biochemicals that deter the herbivore. The effects of this in

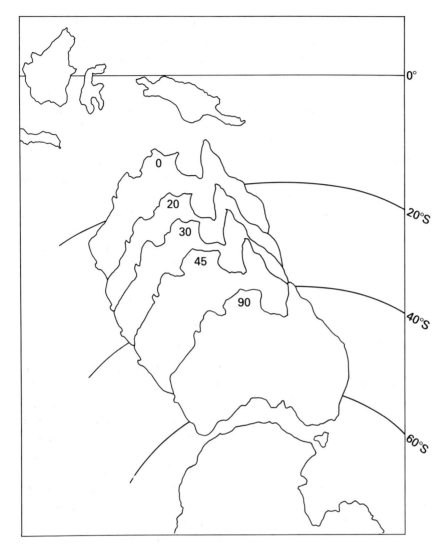

Fig. 8.6 The northward movement of Australia. The most southern position is that before its separation from Antarctica, about 40 million years ago. Positions at 45, 30 and 20 million years ago, and today, are also shown. Most of New Guinea was part of the Australian plate, but Borneo and the other East Indies islands were part of the Asian plate, which Australia progressively approached. After maps in Kemp [17].

depressing the population density of herbivores are shown by the brush opossum, *Trichosurus vulpecula*, whose density in the very different vegetation of New Zealand is five to six times greater than its density in its native Australia [18].

As Australia moved northwards, its northern edge came close to a great oceanic trench (see p. 172), where old ocean crust material was sinking downwards. This lighter material therefore came to underlie the northern edge of the Australian continent. This area (now New Guinea) therefore gradually rose and now consists of high mountains, which were colonized by the Australian rain-forest flora, including *Nothofagus*. This is now the wettest area on earth, though it is close to the driest continent. But, by now, the East Indies had also rotated clockwise towards their present position, so providing a source area from which Asian tropical insects and flowering plants spread eastwards into New Guinea. Asian placental mammals, too, started to spread eastward but, apart, from the aerial bats, only the rats spread naturally as far as Australia, where their 50 species now form 50 per cent of the Australian land mammal fauna. Human beings probably arrived about 30 000 years ago, and the domestic dog, the ancestor of the dingo, about 3500 years ago.

The continental drift history of Australia thus not only explains its climatic changes and so its unusual flora, but also the biotic poverty of the islands to the west of Australia, which are still being colonized — by the Asian biota from the west and by the Australian biota from the east.

New Zealand

The islands of New Zealand have been isolated ever since they split away from Gondwana in the Late Cretaceous or Early Palaeocene, 80–60 million years ago. Though their flora originated as a part of the Antarctic flora, it is now very impoverished, for three reasons. Firstly, much of New Zealand was covered by sea during the Oligocene, greatly reducing its area and therefore its floral diversity. Secondly, during the Pliocene there was a great deal of volcanic activity there, with extensive lava flows. Thirdly, because of its mountainous nature and its far-south position, its flora was decimated by glaciation during the recent Ice Ages. Though much of its old cold-temperate 'Antarctic' flora (including *Nothofagus*) was able to survive this, most of its more warmth-loving flowering plants appear to have died out. This element in the New Zealand flora today has therefore entered the region comparatively recently from Australia: 75 per cent of the New Zealand angiosperm species are also found in that continent. Of the 305 New Zealand angiosperm genera, 38 have arisen within the two islands, but 24 of these have as yet produced only one species each, and none has yet become sufficiently distinct to be recognized as a new, endemic family [19]. New Zealand has no mammals, and therefore does not form part of the system of mammalian biogeographic regions.

South America

The mammal fauna of South America today is characterized by having a few marsupials (opossums) and a diversity of edentate placental mammals that are hardly known elsewhere. Otherwise, it does not seem particularly unlike those of most of the rest of the world, for it includes members of most of the placental orders (see Table 8.1). However, this is largely because of a wave of extinctions and immigrations in the Late Cenozoic, most of which were due to the completion of the Panama Isthmus connecting South America to North America.

At the beginning of the Cenozoic Period, South America already contained early members of the marsupial mammals, an unusual type of ungulate placental mammal, and the edentate placentals [20]. Some of these placentals were able also to disperse to North America in the Late Paleocene [21], probably over an early, short-lived island-chain link. A northward dispersal of other South American animals is documented by the highly unusual fossils of Messel in Germany. In these deposits, laid down in an Early Eocene lake, even the soft parts and stomach contents of the animals have been perfectly preserved [22]. Very surprisingly, the fauna includes a South American ant-eater and two types of typically South American flightless birds, as well as an ancestor of the scaly ant-eater (*Manis*) of the Old World tropics.

The South American marsupials, ungulates and edentates all radiated during the Cenozoic — the marsupials to include the carnivorous borhyaenids as well as the opossums, the ungulates into five herbivorous orders, and the edentates into ant-eaters, armadillos and sloths. This earliest fauna of South American mammals was joined in the Early Oligocene by the ancestors of two other South American groups — the New World monkeys and the caviomorph rodents. The closest living relatives of both these two groups live in Africa. Because the last continuous land link between South America and Africa had broken some 70 million years earlier, in the mid-Cretaceous [23], this situation has provided an interesting biogeographical conundrum that has been vigorously debated for many years [24].

Some workers consider that the African and South American rodents and monkeys evolved independently from earlier, widely spread, more primitive ancestors, and that they entered South America from North America. The problem with this solution is that the rich North American Early Cenozoic mammal faunas contain no possible ancestor for the South American monkeys [25], and no generally acceptable ancestors for the South American caviomorph rodents [26]. Most recent work favours a trans-South Atlantic route [24], perhaps by way of a chain of volcanic islands along the Mid-Atlantic ridge that lay mid-way between northeast Brazil and West Africa in a then-narrower South Atlantic. It seems fair to note that the present state of our knowledge of the northern South Atlantic does not provide evidence for such a chain of islands, but equally does not rule this out [23].

The South American continent in which all these mammals lived consisted of both tropical and subtropical forest, and of open savanna with occasional trees and shrubs. The savanna, in particular, was the environment that permitted the evolution of a great variety of the South American ungulates.

Ever since the beginning of the southern part of the Mid-Atlantic ridge in the Cretaceous, South America had been moving westwards (Fig. 8.7), towards the great oceanic trench in the eastern part of the South Pacific. Eventually, in the Mid-Miocene, it approached the edge of this trench. The western edge of South America therefore came to lie over the zone where Pacific sea floor was being drawn down into the earlier crust. This caused the volcanic and earthquake activity that in turn led to the beginning of the Andean mountain chain. The westward movement of South America led to even more profound geological and biogeographical events further north, for it also caused the final southward extension of the elongate Central American land mass to link the two Americas.

The first biogeographic evidence of the increasing closeness of the two continents came in the Late Miocene and Early Pliocene, when two families from each continent appeared in the other continent, having probably crossed via an island chain. It was soon after this that there took place the great Pliocene uplift of the Andes that doubled their height from 2000 to 4000 m. These lofty mountains now almost completely interrupted the winds that had previously brought moisture-

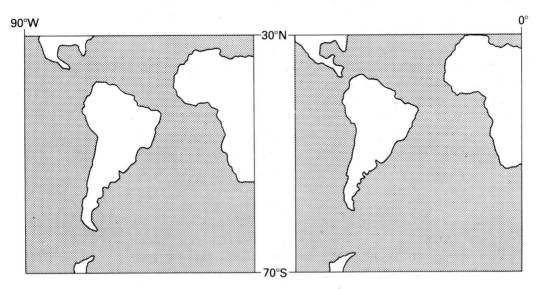

Fig. 8.7 The westward drift of South America. Left, in the Early Oligocene, c. 30 million years ago. Right, in the Late Miocene, c. 10 million years ago. Continental outlines after Smith & Briden [27] and Tarling [28].

laden air eastwards from the Pacific to the high-latitude (30°s) south-eastern parts of the continent. In these areas, the savannas were replaced by treeless pampas, cool steppes, semi-deserts and deserts. As a result, the extensive intermingling of the faunas of North and South America that took place after the final completion of the Panama land bridge also took place at a time of profound ecological change in the South American continent. The fascinating interplay of geology and biogeography in what they have called the Great American Interchange has been analysed in rewarding detail, particularly by the American palaeontologists Larry Marshall and David Webb [29–31].

Each continent had 26 families of land mammal prior to the Inter-change, and about 16 families from each dispersed to the other continent. Of the North American mammals, 29 genera dispersed southwards in the Late Pliocene/early Pleistocene, about 2·5 million years ago. They included shrews, rodents, felids, canids, bears, mastodont elephants, tapirs, horses, peccaries, camels and deer. At the same time, genera of ground-sloth, armadillo (including the giant *Glyptodon*), porcupine and caviomorph rodent dispersed from South America northwards. These were followed northwards, about 1·5 million years ago, by ant-eaters, llamas and opossums.

The fauna that was exchanged between the two continents was one that was adapted to a savanna, open-country environment, suggesting that the environment of the connecting Panama Isthmus must have been of this nature. That inference is supported by types of savanna birds and xerophilous shrubs that now have a disjunct distribution, both north and south of the Isthmus. This distribution presumably arose after the development of the rain forest that is now characteristic of the Panama Isthmus, and which appears to have developed there in the Late Pleistocene.

Though similar numbers of families emigrated northwards and south-wards, the North American emigrants were far more successful than were those from South America. In North America only 29 (21 per cent) of the living land mammal genera are descended from South American stocks, while 85 (50 per cent) of those in South America are derived from North American stocks. Though several other families (such as canids, horses, llamas and peccaries) contributed to the success of North American families in South America, easily the most successful were the cricetid rodents, which diversified into 45 genera in South America.

The difference in the success of the mammals from the two continents may have been partly because the savanna environment was expanding in South America due to the drying effect of the rising Andean Mountains, so providing more ecological opportunity there. But it may also have been because the North American families were the survivors of a long process of Cenozoic evolution and competition throughout the Northern

Hemisphere, while the South American families had been confined to that continent, competing only with one another.

A final phase in the transformation of the South American land mammal fauna came during the Pleistocene. The climatic change that in the Northern Hemisphere caused the Ice Ages and the associated biotic changes (see p. 233 and Chapter 10) also caused a number of extinctions in South America. These included the last of the giant ground-sloths and giant armadillo-like glyptodonts, as well as the horses and mastodont elephants that had immigrated from North America. However, the tapirs and llamas that had originally dispersed from North America became extinct there but survived in South America. As a result these two groups now show a disjunct, relict distribution in which their surviving relatives are found in Asia. Perhaps it is not surprising that few of the South American forms that had dispersed to now-colder North America were able to survive there: only the opossum, armadillo and ant-eater.

The final result of the ecological and biogeographical changes is thus a South American mammal fauna that shows little trace of its original inheritance from the early Cenozoic. The characteristic flowering plant families today include the xerophytic Cactaceae, the Bromeliaceae (the latter include the pineapple, and are often xerophytic), the Tropaeolaceae (including the garden nasturtium) and the Caricaceae (pawpaw tree).

The Old World tropics — Africa, India and South-East Asia

Three original land masses contributed to this area, and all were originally part of Gondwana (see Fig. 7.5). South-East Asia may have been the first to unite with Asia, some time in the Triassic, while India collided with southern Asia in the Early Eocene, its continuing northwards movement throwing up the Himalaya Mountains and raising the Tibetan Plateau later in the Cenozoic. Africa was therefore the last of these three to become united with the northern continents, joining them during the Late Cenozoic. But it had never lain far from southern Eurasia, and probably acted as a giant stepping-stone for the dispersal of the tropical flowering plants from their origin in the South American/African tropics to the tropics of southern Asia. Many elements of the tropical biota doubtless became widespread throughout the region and, before the Late Cenozoic cooling of the Northern Hemisphere, would also have ranged northwards into higher latitudes of Eurasia. However, that cooling, together with the spread of seas and deserts in the Middle East, led to a new division of the Old World tropics into a western division, made up of Africa alone, and an eastern 'Oriental' section made up of India and South-East Asia.

It is not yet possible to trace in full the separate contributions of

Africa, India and South-East Asia to the final Old World tropical biota. As always, the fossil record of mammals is easier to interpret, and can then be used as a guide for the reconstruction of the histories of the angiosperm floras. However, there was probably one significant difference between these histories. Because flowering plants had evolved and diversified in the Mid-Cretaceous, and because they have good powers of dispersal, many families probably became widespread throughout Africa and Eurasia. Most of the early types of placental mammal, on the other hand, evolved in the Northern Hemisphere at the beginning of the Cenozoic, and therefore spread southwards into Africa and India as and when that became possible, in the Paleocene and Eocene.

The best-known and most distinctive of the mammal faunas is that of Africa. The shallow seas that separated Africa from Eurasia (see Fig. 8.8a) in the early Cenozoic were not the only barriers to biotic exchange, for northern Africa lay in the northern Horse Latitude arid belt. Nevertheless, some placentals managed to enter Africa from the north: an invasion by early insectivorans may have taken place as early as the Cretaceous–Cenozoic boundary, and there was also a later dispersal, involving early primates and creodonts, which may have occurred near the Paleocene–Eocene boundary [32]. A more complete African mammal record only begins in the Late Eocene to Early Oligocene faunas of northern Africa [33], which contain two slightly different faunal groups. One group consists of the elephants, hyracoids (conies), aquatic sirenians (sea-cows) and the extinct embrithopods. These were all the descendants of an ungulate stock that had entered Africa some time earlier, in the Paleocene or Early Eocene, and had since diversified into these four unique orders, all endemic to Africa. The other group of early African placentals had apparently entered that continent more recently; this included artiodactyls, creodonts, insectivores, rodents, and also the earliest members of the anthropoid primate line (which later evolved into apes and man). Marsupials, which had reached the European part of Euramerica during the early Cenozoic via Greenland (see p. 230), are also known to have been present in northern Africa by the Early Oligocene [34]. Later in the Cenozoic, other new endemic African orders of placental mammal evolved — the elephant shrews and Cape golden moles.

The Late Cenozoic geographical and biogeographical relationship between Africa and Eurasia is quite complicated. The first collision between the two continents took place about 19 million years ago, in the Early Miocene, and was between the Arabian and Turkish regions of the continents (Fig. 8.8b). The closure of the seaway between them interrupted the circum-equatorial world oceanic circulation, and this in turn may have been the cause of the climatic deterioration that took place in central Europe at that time. Carnivorans, suids (pigs), bovids

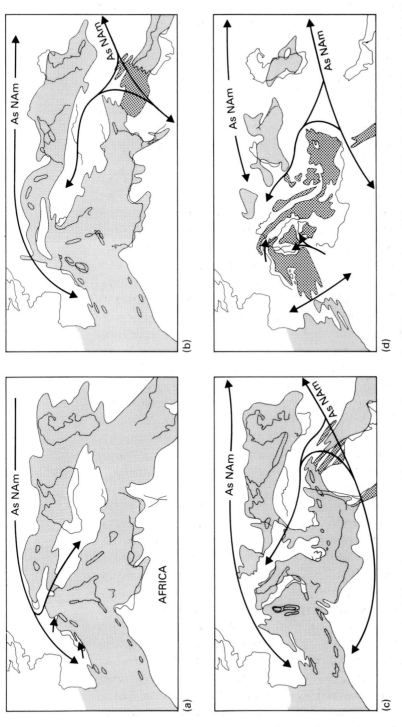

Fig. 8.8 Reconstructions of the Mediterranean area: (a) in the Late Oligocene/Early Miocene boundary; (b) in the middle of the Early Miocene; (c) the beginning of the Late Miocene; (d) near the end of the Late Miocene. Light tint: sea; dark tint, evaporitic deposits laid down as the seas dried up. Arrows show directions of mammal dispersals; As, Asia; NAm, North America. Some present-day geographical outlines have been added to aid recognition and location. After Steininger *et al.* [35].

(cattle, antelope, etc.) and cricetid rodents passed across the new land bridge from Asia to Africa, while elephants, creodonts and primates passed in the opposite direction.

There was a re-opening of the seaway in the Mid-Miocene, 16 million years ago, which led to a warmer, moister climate through central Europe. But this marine link was not wide enough, or not constant enough, to prevent the dispersal to Eurasia of new types of primate, elephant and suid that had evolved in Africa. The seaway was finally broken near the beginning of the Late Miocene, 12 million years ago, by rising mountains in Arabia, Turkey and the Middle East, when the early horse *Hipparion*, which had evolved in North America, appears in both Eurasia and Africa, while rhinoceros, hyenas and sabre-toothed cats dispersed from Africa to Eurasia (Fig. 8.8c).

A final, dramatic event was the closure of the western connection between the Mediterranean and the Atlantic near the end of the Late Miocene, six million years ago, due to a world-wide fall in sea levels as well as to the rise of mountains in both Spain and north-west Africa. Much, perhaps all, of the Mediterranean Sea then dried up completely, leaving an immense plain 3000 m below sea level, covered with a thick deposit of rock salt (Fig. 8.8d). The resulting drier climate around the Mediterranean led to the expansion of sclerophyll evergreen woodlands, and may also have led to the evolution of the typical Saharan flora further south in Africa. This drier European environment received many Asian mammals that inhabited steppe and savanna conditions, while the African hippopotamuses spread to its river systems, and African rhinoceroses lived on its plains. It was not until the Pliocene that the Atlantic gateway re-opened, the Mediterranean was re-filled, and the climate of Europe became warm and humid.

Though India, because of its geographic origin in southern Gondwana, might have contributed a distinctive element to the tropical biota of Eurasia when it joined that continent, there is little sign of it. This is probably because India separated from Gondwana in the Early Cretaceous, before placentals had evolved and when the flowering plants had only just appeared, and so may at first have lacked these groups altogether. Some flowering plants doubtless dispersed to India as it drifted northwards through the Indian Ocean, and it may have acted as a giant stepping-stone for some dispersal between Australia and Madagascar [36]. India certainly had a varied angiosperm flora in the Early Eocene [37], soon after it became united with Asia. Apart from the presence of the Proteaceae (some of which have dandelion-like wind-dispersed seeds), there is little evidence of Gondwana angiosperms in this Early Eocene flora, and it seems more likely that it dispersed to India from Asia. The Indian flora today is dominated by flowering plants found also in South-East

Asia, and it contains no endemic families [38]. Almost certainly, India originally had early types of Mesozoic mammal, of a more primitive evolutionary grade than the marsupials and placentals, as these primitive types are widespread in the Mesozoic world. Unfortunately, the mammal fauna of India in the Cretaceous and Paleocene, before it collided with Asia, is unknown. Such primitive Indian mammals would soon have become extinct in the face of the competition from the Asian placentals that would have entered India after the collision, in a wave of extinction rather like that which followed the connection between the Americas in the Pliocene (see above).

The results of the Miocene interchange of mammals between Africa and tropical Eurasia can still be seen today. Primitive primates such as the lemurs and lorises, as well as Old World monkeys and porcupines, apes, rhinoceroses, elephants, and the pangolin (*Manis*), are all found exclusively in these two areas. But in nearly every case each group is represented by different genera in the two areas. For example, the African rhinoceros, elephant and porcupine all belong to genera different from those found in the Oriental region. Similarly, the lemurs of Madagascar and the chimpanzees and gorillas of Africa are not found in the Oriental region, where these groups are represented by the lorises and by the orang-utan and gibbon.

Though these differences might have developed in any case, because of the great distance between Africa and tropical Eurasia, these two faunas also became isolated from one another by climatic and geographic events in the Middle East. Another reason for the isolation of the African and tropical Eurasian faunas from one another was that the direct route between them was interrupted by the development of the Red Sea in the Pliocene, and by the extension of deserts in the Middle East. The two faunas also became more different from one another because of the uplift of East Africa in the Late Miocene, and the consequent increased dryness of that region. Together with the general drying of the world climate at that time, this reduced the eastward extent of the great belt of African rain forest, which now became restricted to West Africa and the Congo Basin. This decrease in the extent of the African tropical flora is probably the reason why it is much less diverse than that of South America and South-East Asia. East Africa instead became covered by woodland and bushland with a ground cover of herbs and grass. (The replacement of this environment by the dry grasslands called savanna is probably a result of human activities — overgrazing, cultivation and clearance by fire [39].) The huge herds of browsing and grazing ungulates that now live in that area, such as the many types of antelope (impala, gazelle, gnu, and others), giraffes, buffalo, zebra and wart-hogs, are now thought of as the 'typical' fauna of Africa.

But in reality these are late-comers to the African scene; their ancestors are not known in Africa until the Middle Miocene. Our own genus, *Homo*, also seems to have evolved in this environment (see p. 286).

A similar increase in aridity, related to the uplift of the Himalaya mountains to new heights, affected the northern parts of the Indian subcontinent about three million years ago, and this led to an increase in the numbers of grazing mammals such as horses, antelope and camels, as well as of elephants [40].

No such climatic changes affected the tropical forests of South-East Asia, which contain a great diversity of primitive families of flowering plant. This fact led the Russian botanist Armen Takhtajan to suggest [41] that the area was the original home in which flowering plants had evolved. However, it now seems more likely that the diversity is the result of a comparatively recent fusion of two separate angiosperm floras. One of these may have been the original tropical angiosperm flora of southern Asia, while the other is of Gondwanan origin, having dispersed into South-East Asia from India or Australia. But, as always, the nature of the fossil record of angiosperms (see above) makes it difficult to document clearly the historical biogeography of these floras. This same difficulty lies in the way of solving another plant geographic problem in the Old World tropics. The Miocene exchange of mammals between Africa and the Oriental region was doubtless accompanied by a similar exchange of flowering plants. But the inadequate fossil record makes it at present impossible to distinguish which of the angiosperm families that are found in both areas are merely part of a common Late Cretaceous or Early Cenozoic inheritance, and which may have dispersed between the two areas later in the Cenozoic.

At the margins of Africa, two areas contain biota that merit special consideration: the fauna of Madagascar, and the flora of the Cape region.

Madagascar

This large (640 000 km²) island was heavily forested until recently (see Fig. 4.9). Its mammal fauna is notable for two features: most of the families are endemic to the island, and most of them seem to be ancient, primitive offshoots from their respective Orders. The Primates comprise three endemic families of lemur; the Insectivora comprise an endemic family, the tenrecs; the Rodentia include an endemic subfamily of cricetine (as well as the rats and mice introduced by humans), and the Carnivora comprise a number of genera of viverrid (civet) that may all belong to a single endemic family. The only other land mammals that have reached the islands naturally are a pygmy hippopotamus that became extinct during the Pleistocene, and a river-hog.

These features of the mammal fauna are paralleled in other groups of animals and plants, and suggest a long independent history for the island. This is confirmed by the recent demonstration, from the history of widening of the Mozambique Channel that separates Madagascar from Africa, that the island parted from the mainland in the Middle Jurassic and moved southwards to reach its present position in the Early Cretaceous.

It would be natural to deduce from these facts that the primitive members of the four mammalian orders entered Madagascar when the island was closer to the mainland, and that it was its continued movement away from the mainland that made it impossible for later types of mammal to reach the island. However, this cannot have been the reason if Madagascar was already in its present position in the Early Cretaceous, long before any placental mammals evolved. Any change in the ease of crossing of the Mozambique Channel must therefore have been the result of changes in the currents in the Channel, or the disappearance of intervening volcanic islands similar to the Comoro Islands that lie between north-western Madagascar and the mainland.

The Cape flora

A surprising feature of the plant geography of the Old World tropics is the recognition of a small area at the southern tip of Africa as a separate 'Cape' floral region (Fig. 8.3). This recognition is based on the fact that the region contains six endemic angiosperm families, together with an extremely high rate of generic endemism (19.5%) and a great richness at species level — it contains over 8500 flowering plant species. The vegetation, known as 'fynbos', consists of fine-leaved, bushy, sclerophyll plants, and is dominated by members of the Restionaceae in particular and also of the Ericaceae and Proteaceae.

The origins of the flora are still unclear, but there are several significant facts [42]. Firstly, though they exist at present in a Mediterranean-type climate with dry summers, many of the Cape angiosperms show a pattern of rapid vegetative growth at that time, suggesting that they were originally adapted to a more temperate climate with summer rainfall. Secondly, the flora of some mountains in tropical Africa includes about 225 species found in the Cape flora, as well as other species related to Cape species but apparently more primitive than the latter. Thirdly, the climate of the Cape region appears to have changed from temperate to Mediterranean within the last few million years [43].

Taken together, these facts suggest that some elements, at least, of the Cape flora were originally found in temperate regions further north in Africa. As the climate of that region became hotter, their area of distribution changed: some moved to higher altitudes, while others either moved southwards or became restricted to southern Africa. However, it is uncertain whether all the elements of the Cape flora had such an origin and, in any case, such a uniform history is unlikely. For example, though the nearly 300 species of the Restionaceae dominate the Cape flora, there is no sign of that family, living or fossil, further north on the mainland of Africa. Their diversity in the Cape region suggests that they may have inhabited that region for a considerable time, rather than being recent immigrants. The family is also known in Madagascar (one species), Australia and South America, suggesting that it may show a relict distribution within Gondwana. Finally, some elements of the Cape flora, such as the Gunneraceae and the tree *Metrosideros* (known also in Hawaii and many Pacific islands, see p. 156) may have dispersed to the region across the sea. So, like most floras, that of the Cape region is a complex of elements of varying origins, varying histories, and varying times of arrival in the area.

The Northern Hemisphere — Holarctic mammals and Boreal plants

In contrast to the complexity of the geological and geographical history of the Southern Hemisphere, that of the Northern Hemisphere has been far more uniform. Though shallow epicontinental seas, and the developing North Atlantic, have from time to time subdivided the land areas of North America and Eurasia into different patterns, the two continents have never been far apart, so that dispersal between them has usually been fairly easy. Furthermore, in the recent past their faunas and floras have all suffered from the severe climatic effects of the Ice Ages, so that the most fundamental aspect of the biogeography of these two regions now is that they are the great regions of temperate- and cold-adapted biota. Though the two continents are usually distinguished as being separate 'Nearctic' and 'Palaearctic' zoogeographic regions, they are sometimes considered as a single 'Holarctic' region, like the single 'Boreal' region of plant geographers.

Though Greenland may have been separated from North America by a sea channel for a short time at the end of the Cretaceous, there was otherwise a continuous connection from North America to Europe via Greenland until the end of the Eocene. Sea levels were then lower, so that the European continental shelf was dry land, and all such present-day islands as the Faroes, Orkneys, Great Britain and Spitzbergen were merely a part of a more extensive European continent. Greenland was connected to this by two different routes [44] (Fig. 8.9). The Thulean route was along a now-submerged ridge of land that linked eastern Greeland via Iceland to Europe until the end of the Early Eocene. The de Geer route ran from northern Greenland to the most extreme north-westerly corner of the European continent. This connection continued until Greenland finally separated from Europe at the end of the Eocene

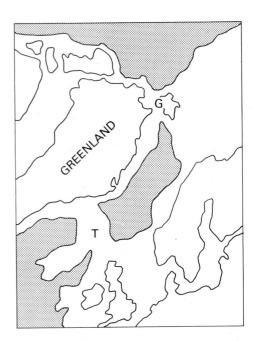

Fig. 8.9 Connections between North America and Europe in the Eocene. G, de Geer route; T, Thulean route (see text). After Tarling [45].

but, because of its high northern latitude, must have been a climatic filter. Much of the early evolution of placentals appears to have taken place in this Euramerican continent, in which there is a rich series of Late Cretaceous and Early Cenozoic deposits containing fossil mammals. The primates, rodents, bats, artiodactyls, perissodactyls, Carnivora and modern insectivores are all known first from that area [46]. Though the European Cenozoic fauna is not known until the Late Paleocene, it is then very similar to that of North America; nearly all the European families are also known in North America, and some genera are found in both continents. On the other hand a number of North American Late Paleocene families are unknown in Europe, and the climate of the northern connection between the two continents appears to have acted as a filter. Asia during this time was isolated from both Europe and North America. Marsupials, similar to the living South American opossum, are unknown from Asia, but are known to have reached Europe from North America by the Early Eocene, and survived in both those continents until the Miocene.

The second pattern of relationship between the northern continents began in the Early Oligocene. The North Atlantic now separated Europe from North America, and such new European groups as palaeothere horses and primitive relatives of the camel could not cross to North America. On the other hand, the drying-up of the Obik Sea, perhaps because more water was now becoming locked up in the spreading ice-sheets of the Antarctic [47], now allowed the Asian fauna direct entry into Europe. At the same time the climate was warmer than at the end of Eocene, so that a greater variety of mammals was able to enter Asia from North America via the Bering connection.

From now on, that Bering link was the only route between Siberia and Alaska. Because this route lies at a high northern latitude, the climatic changes of the Late Cenozoic not only affected the nature of the biota of the two northern continents, but also affected the nature of the biotic exchange between them. The climatic deterioration that began in the Early Oligocene therefore not only steadily reduced the area occupied by megathermal plants, but also the floral exchange via the cooling Bering region (though the North Pacific was bordered by continuous broad-leaved deciduous forests until at least the Middle Miocene). It also led to the expansion of the Northern Hemisphere gymnosperm family Pinaceae — pine, fir, spruce and larch. It was at one time thought that the climatic cooling caused a wholesale southward movement, through the whole of the Northern Hemisphere, of an 'Arcto-Tertiary flora' that had evolved in the Arctic during the Cretaceous and had survived until today, little changed, in south-eastern North America and east-central Asia. However, this concept has not been supported by later knowledge of the floral history of the Alaskan region. Instead, the Northern Hemisphere angiosperm floras appear to have adapted to the Late Cenozoic climatic change in three ways: by the adaptation of some genera to changed, cooler climates; by the restriction of the range of some genera and their replacement by other already-existing genera that preferred a cooler climate; and by the evolution of new genera that preferred those cooler climates.

Much of the Late Cenozoic microthermal vegetation of the Northern Hemisphere appears to have evolved *in situ* from ancestors within the same area. Since there was little exchange of plants between North America and Eurasia during this time, these two floras steadily diverged. Though it has also been suggested that they shared a common 'Madro-Tertiary' dry flora, exchanged via a low- to middle-latitude dry corridor, this concept also now seems erroneous [48]. Instead, the ancestors of the plants found today in the dry region of south-western North America and the Mediterranean also appear to have lived in forests with moderate or high rainfall, and they have, in each region, become adapted independently to a drier environment.

The crucial nature of the climate of the Bering region can also be seen in its influence upon faunal exchange between North America and Eurasia. When the climate became cool, as in the Early Oligocene, few mammals crossed. When it improved again a little later in the Oligocene, a number of Asian mammals dispersed to North America; some of these had evolved within Eurasia, while others had dispersed to that continent from Africa. The final climatic deterioration in the Bering region began in Miocene times, perhaps related to the increase in Antarctic glaciation (see p. 199). From then on, most of the mammals that dispersed were large forms and, even more significantly, types that are tolerant of cooler temperatures — such warmth-loving forms as apes and giraffes could not reach North America. This climatically based exclusion became progressively more restrictive, until in the Pleistocene only such hardy forms as the mammoth, bison, mountain sheep, mountain goat, musk-ox and human beings were able to cross. The final break between Siberia and Alaska took place 13 000 to 14 000 years ago.

Despite the long history of intermittent connection between the Nearctic and Palaearctic regions, each has certain groups of animals that have never existed in the other one, and also groups that did reach both regions, but became extinct in one and were never replaced by subsequent colonization from the other. Pronghorn antelopes, pocket gophers and pocket mice, and sewellels (the last three groups are all rodents) are unknown in the Palaearctic region, whereas hedgehogs, wild pigs and murid rodents (typical mice and rats) are absent from the Nearctic region. The domestic pig has been introduced to North America, as have mice and rats at various times. The horse became extinct in the Americas during the Pleistocene, but had crossed the Bering connection to Eurasia. Horses were therefore unknown to the American Indians until they were introduced by the Spanish *conquistadores* in the sixteenth century.

Though the Ice Ages did not commence until the end of the Pliocene, the steadily cooling climates had already exerted a great influence upon the floras of the northern continents. For example, there was a considerable change in the European flora during the Pliocene. Only 10 per cent of the Early Pliocene flora of Europe still survives there, contrasted with over 60 per cent of its Late Pliocene flora. The intervening three million years had therefore seen a drastic modernization of the flora of Europe, as it started to adapt to climatic changes that now became greatly exaggerated as the Pleistocene Ice Ages commenced. These Ice Ages stripped North America and Eurasia of virtually all tropical and sub-tropical animals and plants. This happened so recently that the faunas and floras have as yet had no time to develop any new, characteristic groups. Since they also have no old relict groups, such as the marsupials, it is the poverty and the hardiness of their faunas that distinguishes them from those of other regions. Many groups of animals are absent altogether and, of the groups that are present, only the more hardy members have been able to survive. Even these become progressively fewer towards the colder, Arctic latitudes. In North America there is, in addition, a similar thinning-out of the fauna in the higher, colder zones of the Rocky Mountains. This is a general feature of the fauna and flora of high mountains, as described in Chapter 2.

The Palaearctic fauna was almost completely isolated from the warmer lands to the south by the Himalayas and by the deserts of North Africa and southern Asia, and has therefore received hardly any infiltrators to add variety. The situation was, eventually, rather different in the Western Hemisphere. During the Early Cenozoic there had been hardly any exchange of either animals or plants between North and South America, presumably because there was still a wide ocean gap between the two continents. But the Panama Isthmus steadily extended southwards, forming a lengthening tropical extension of North America. It therefore provided a region within which North American mammals adapted to tropical conditions could survive or evolve. Many of these dispersed to South America when the final southern link to that continent was established (see p. 222), whereas North America was successfully colonized by only three types of South American mammal (opossums, armadillos and ant-eaters), along with a number of birds such as humming-birds, mocking-birds and New World vultures.

For plants, for some reason, the situation was reversed. Instead of surviving in the Panama Isthmus, few of the North American mega-thermal plants survived the Late Cretaceous climatic cooling. The bulk of the lowland vegetation of Central America is, instead, of South American origin, perhaps because that great tropical region had produced an enormous variety of tropical plants. The mesothermal plants of North America were more successful at colonizing South America, presumably using the cooler mountainous spine of Central America as their route from the Rockies to the Andes.

References

1 Miller, H.A. (1982) Bryophyte evolution and geography. *Biological Journal of the Linnean Society* **18**,145–196.
2 Keast, A. (1973) Contemporary biotas and the separation sequence of the southern continents. In *Implications of Continental Drift to the Earth Sciences*, Vol. 1 (eds D.H. Tarling & S.K. Runcorn). Academic Press, New York and London, pp. 309–343.
3 Gressitt, J.L. (1974) Insect biogeography. *Annual Review of Entomology* **19**, 293–321.
4 Cracraft, J. (1974) Continental drift and vertebrate distribution. *Annual Review of Ecology and Systematics* **5**,215–261.
5 Savage, J.M. (1973) The geographic distribution of frogs: patterns and predictions. In *Evolutionary Biology of the Anurans* (ed. J.L. Vial). University of Missouri Press, Columbia, pp. 351–445.
6 Cracraft, J. (1973) Continental drift, palaeoclimatology, and the evolution and biogeography of birds. *Zoological Journal, London* **169**,455–545.
7 Eisenberg, J.F. (1981) *The Mammalian Radiations. An Analysis of Trends in Evolution, Adaptation, and Behavior.* University of Chicago Press, Chicago.
8 Flannery, T. (1991) The mystery of the Meganesian meat-eaters. *Australian Natural History* **23**,722–729.

9 Takhtajan, A. (1986) *Floristic Regions of the World*. University of California Press, Berkeley.

10 Heywood, V.G. (1978) *Flowering Plants of the World*. Oxford University Press, Oxford.

11 Wolfe, J.A. & Upchurch, G.R. (1987) North American nonmarine climate and vegetation during the Late Cretaceous. *Palaeocology, Palaeoclimatology and Palaeoecology* **61**,33−77.

12 Muller, J. (1981) Fossil pollen records of extinct angiosperms. *Biological Reviews* **47**,1−142.

13 Keast, A. (1981) *Ecological Biogeography of Australia*, Vols I−III. Junk, The Hague.

14 Gressitt, J.L. (ed.) (1982) *Biogeography and Ecology of New Guinea*, Vols I, II. Junk, The Hague.

15 Whitmore, T.C. (ed.) (1987) *Biogeographical Evolution of the Malay Archipelago*. Oxford Monographs in Biogeography, 4.

16 Beadle, N.C.W. (1981) Origins of the Australian flora. In *Ecological Biogeography of Australia* (ed. A. Keast), Vols I, II. Junk, The Hague, pp. 407−426.

17 Kemp, E.M. (1981) Tertiary palaegeography and the evolution of Australian climate. In *Ecological Biogeography of Australia* (ed. A. Keast), Vols I, II. Junk, The Hague, pp. 33−80.

18 Tyndale-Biscoe, C.H. (1979) Ecology of small marsupials. In *Ecology of Small Mammals* (ed. D.M. Stoddart). Chapman & Hall, London, pp. 342−379.

19 Godley, E.J. (1975) Flora and vegetation. In *Biogeography and Ecology in New Zealand* (ed. C. Kuschel). Junk, Amsterdam, pp. 177−229.

20 McKenna, M.C. (1980) Early history and biogeography of South America's extinct land mammals. In *Evolutionary Biology of the New World Monkeys and Continental Drift* (eds R.L. Ciochon & A.B. Chiarelli). Plenum Press, New York and London, pp. 43−77.

21 Gingerich, P.D. (1985) South American mammals in the Paleocene of North America. In *The Great American Interchange* (eds F.G. Stehli & S.D. Webb). Plenum, New York, pp. 123−137.

22 Storch, G. (1993) 'Grube Messel' and African−South American faunal connections. In *The Africa−South America Connection* (eds W. George & R. Lavocat). Oxford Monographs in Biogeography No. 7, pp. 76−86.

23 Parrish, J.T. (1993) The palaeogeography of the opening South Atlantic. In *The Africa−South America Connection* (eds W. George & R. Lavocat). Oxford Monographs in Biogeography No. 7, pp. 8−27.

24 George, W. & Lavocat, R. (eds) (1992) *The Africa−South America Connection*. Oxford Monographs in Biogeography, 7.

25 Aiello, L.C. (1993) The origin of the New World monkeys. In *The Africa-South America Connection* (eds W. George & R. Lavocat). Oxford Monographs in Biogeography No. 7, pp. 100−118.

26 George, W. (1993) The strange rodents of Africa and South America. In *The Africa−South America Connection* (eds W. George & R. Lavocat) Oxford Monographs in Biogeography No. 7, pp. 119−141.

27 Smith, A.C. & Briden, J.C. (1977) *Mesozoic and Cenozoic Continental Maps*. Cambridge University Press, Cambridge.

28 Tarling, D.H. (1980) The geologic evolution of South America with special reference to the last 200 million years. In *Evolutionary Biology of the New World Monkeys and Continental Drift* (eds R.L. Ciochon and A.B. Chiarelli). Plenum Press, New York and London, pp. 1−41.

29 Marshall, L.G. (1981) The Great American Interchange − an invasion-induced crisis for South American mammals. In *Third Spring Systematic Symposium: Crises in Ecological and Evolutionary Time* (ed. M.H. Nitecki). Academic Press, New York and London, pp. 133−229.

30 Marshall, L.G. *et al.* (1982) Mammalian evolution and the Great American Interchange. *Science* **215**,1351–1357.

31 Webb, S.D. (1985) Late Cenozoic mammal dispersals between the Americas. In *The Great American Biotic Interchange* (eds F.G. Stehli & S.D. Webb). Plenum, New York, pp. 357–386.

32 Gheerbrant, E. (1990) On the early biogeographical history of the African placentals. *Historical Biology* **4**,107–116.

33 Coryndon, S.C. & Savage, R.J.G. (1973) The origin and affinities of African mammal faunas. *Special Paper in Palaeontology* **12**,121–135.

34 Brown, T.M. & Simons, E.I. (1984) First record of marsupials from the Oligocene in Africa. *Nature* **308**,447–449.

35 Steininger, F.F., Rabeder, G. & Rögl, F. (1985) Land mammal distribution in the Mediterranean Neogene: a consequence of geokinematic and climatic events. In *Geological Evolution of the Mediterranean Basin* (eds D.J. Stanley & F.C. Wezel). Springer, New York and Berlin, pp. 559–571.

36 Schuster, R.M. (1976) Plate tectonics and its bearing on the geographical origin and dispersal of angiosperms. In *Origin and Early Evolution of Angiosperms* (ed. C.B. Beck). Columbia University Press, New York and London, pp. 48–138.

37 Lakhanpal, R.N. (1970) Tertiary floras of India and their bearing on the historical geology of the region. *Taxon* **19**,675–694.

38 Mani, M.S. (ed.) (1974) *Ecology and Biogeography in India. Monographiae Biologiae* **23**,1–773. Junk, The Hague.

39 Andrews, P. & Van Couvering, J.A.H. (1975) Palaeoenvironments in the East African Miocene. *Contributions to Primatology* **5**,62–103.

40 Singh, G. (1988) History of aridland vegetation and global climate. *Biological Reviews* **63**,156–190.

41 Takhtajan, A. (1969) *Flowering Plants, Origin and Dispersal.* Oliver & Boyd, London.

42 Taylor, H.C. (1978) Capensis. In *Biogeography and Ecology of Southern Africa* (ed. M.J.A. Werger). Junk, Amsterdam, pp. 171–229.

43 Cowling, R. (ed.) (1992) *The Ecology of Fynbos. Nutrients, Fire and Diversity.* Oxford University Press, Oxford.

44 McKenna, M.C. (1983) Cenozoic paleogeography of North Atlantic land bridges. In *Structure and Development of the Greenland–Scotland Ridge* (eds M.H.P. Bott *et al.*). Plenum, New York and London, pp. 351–399.

45 Tarling, D.H. (1982) Land bridges and plate tectonics. *Mémoir Spéciale Geobios* **6**,361–374.

46 Cox, C.B. (1974) Vertebrate palaeodistributional patterns and continental drift. *Journal of Biogeography* **1**,75–94.

47 Zachos, J.C., Breza, J.R. & Wise, S.W. (1992) Early Oligocene ice-sheet expansion in Antarctica. *Geology* **20**,569–77.

48 Wolfe, J.A. (1975) Some aspects of plant geography of the Northern Hemisphere during the Late Cretaceous and Tertiary. *Annals of the Missouri Botanical Garden* **62**,264–269.

9 *Interpreting the past*

Although it is obvious that there is a continuous spectrum of biogeographical problems from the most recent to the most ancient, many workers distinguish between two different approaches — ecological biogeography and historical biogeography [1]. (Of course, there is an overlap between these two [2]; for example, the biogeographic problems arising from the Ice Ages might be considered either as ecological, if they concern the Pleistocene ancestors of living groups, or as historical if they instead concern organisms that are less well-understood because they have left no living descendants.)

Because it is concerned with the more distant past, historical biogeography has been greatly affected by the 'revolution in the earth sciences' that resulted from the acceptance of the theory of plate tectonics. Previously, it had been assumed that the gaps in the distribution of organisms represented inhospitable areas that had existed before the organisms had spread. They had therefore been obliged to disperse actively across such barriers, which had thereafter isolated them as separate units, within each of which independent genetic change had led to divergence into distinct species — and perhaps ultimately into distinct genera, etc., as these differences became more numerous, comprehensive and fundamental.

But plate tectonic theory suggested a greater role for geographical changes in subdividing and isolating populations of organisms. Whole continents could split, each fragment carrying away its cargo of both living organisms and buried fossils — acting, as the American palaeontologist Malcolm McKenna has so vividly described them [3] both as arks carrying living organisms and as Viking funeral ships carrying the dead (fossils). In this new approach, known as 'vicariance biogeography', the emphasis is on new barriers, subdividing a previously continuous range of distribution, rather than on species dispersing across a pre-existing barrier as in the older 'dispersal' approach.

As sometimes happens in science, the supporters of these two approaches became antagonistic, and the argument became polarized. Much biogeographical literature from the 1960s (especially that of the pan-biogeographical group, see below) contains distasteful and unnecessary attacks on other biogeographers extending as far back as Charles Darwin. However, it is obvious that both phenomena not only exist, but also can take place simultaneously. For example, the completion of the Panama land bridge allowed dispersal of some northern mammals southwards into South America at the same time as it provided a vicariance event in dividing a previously single population of marine organisms into separate Caribbean and East Pacific populations.

Because the event in question usually took place during past geological time, the only direct evidence is historical — either geological or palaeontological. If it is possible to demonstrate a clear correlation between a geological or climatic event and the subsequent divergence of biota on the appropriate geographical units, that is strong evidence for vicariance. For example, as Malcolm McKenna has shown [4], the mammal fauna of North America and that of Europe were very similar until the Middle Eocene. Geological and geophysical research suggests that the North Atlantic finally separated these two areas at just that time. But this evidence is nevertheless circumstantial. The absence of data supporting one explanation does not prove that it did not take place, and the same is true of the alternative explanation. Except for comparatively recent events, the evidence in most cases is not adequate to allow us to make a firm decision between dispersal and vicariance explanations.

On the other hand, though this may be true of the individual case viewed in isolation, there is a fundamental difference between the likely results of the two processes when a number of examples of disjunct distributions are considered together. A vicariant event, caused by tectonic changes or by the spread of shallow seaways, will simultaneously affect a considerable variety of organisms, whose subsequent evolutionary history is likely to be altered as a result of the ensuing isolation. Any such parallelism in apparent biogeographic history is therefore good *prima-facie* evidence for vicariance. Dispersal, on the other hand, is by its very nature a rare and comparatively isolated event, which will only very occasionally affect more than one type of organism — as, for example, when a violent cyclone or tidal wave carries a number of organisms from one area to another. Dispersal is therefore likely to appear only as the occasional aberrant event in the biogeographical record.

The realization of this difference has to a large extent calmed the argument between the proponents of dispersalism and of vicarianism. Even though several of the current techniques of analysis of historical biogeography originally arose from a primarily vicariance approach, the data are examined without any preconceived ideas, and judgement as to the original cause of the disjunction is made subsequently, on the basis outlined above.

The current vicariance theories can be divided into two groups, one of which relies primarily on 'cladistic' analysis of the characteristics of the organisms involved, while the other instead commences with analysis of the patterns of distribution. The cladistic method originated as a new approach to taxonomy, propounded by the German worker Willi Hennig in 1950. This treated the process of evolutionary change as a series of branching events, or 'dichotomies', at each of which a single group divides into two daughter groups. At each dichotomy, one or more of the characteristics of the group changes from a primitive or 'plesiomorphic' state into a derived or 'apomorphic' state. The evolutionary history of the group can then be portrayed as a branching 'cladogram'. So, in Fig. 9.1, characters a–g evolved after the divergence between group 1 and groups 2–6. They are therefore derived or apomorphic relative to the characters of group 1 (in which these characters have remained primitive), but primitive or plesiomorphic for groups 2–6. Other new, apomorphic characters h–y evolved at different points within the evolutionary history of groups 2–6, and can therefore be used to analyze their patterns of relationship.

As far as possible, it is assumed that each apomorphic evolutionary event only occurred once in the history of each group of related taxa (a concept known as economy of hypothesis, or 'parsimony'), and the taxa are arranged on the cladogram in such a way as to minimize the number of parallelisms. For example in Fig. 9.1 it is most parsimonious to believe that character h has evolved twice, because that involves the assumption of only that single additional evolutionary event. The alternative is to transfer the origin of group 2 to near the base of groups 3/4, with the consequent need to assume that characters i–k had been lost in group 2 — an assumption of three additional evolutionary events instead of only one.

Phylogenetic biogeography

The potential of this method as a tool for biogeographic analysis was first realized by the Swedish entomologist Lars Brundin, who in 1966

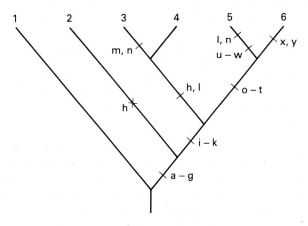

Fig. 9.1 Cladogram of the relationships between six groups, using characteristics a–y. See text for explanation.

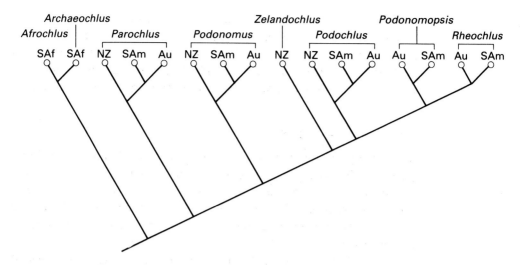

Fig. 9.2 Simplified taxon–area cladogram of some of the Gondwana genera of podonominine chironomid midge studied by Brundin [5]. The names in italics are those of the genera involved, while the circles represent individual species. The initials indicate the continent in which each species is found: Au, Australia; NZ, New Zealand; SAf, South Africa; SAm, South America. The African genera appear to have diverged first. In each of the other genera, the divergence of the New Zealand species preceded the divergence between the South American and the Australian species.

analysed the distribution of chironomid midges belonging to three sub-families found in the Southern Hemisphere. He first produced a cladogram of the evolutionary relationships of the species. In place of the name of each species he then instead inserted the name of the continent on which it is found, transforming the phyletic cladogram into a taxon–area cladogram (Fig. 9.2). The result was a consistent pattern, in which the African species appeared to have diverged first, followed in turn by those of New Zealand, South America and Australia [5]. This sequence, based upon the evolutionary relationships of the midges, was independently supported by geophysical data on the sequence of break-up of the Gondwana supercontinent. (India and Antarctica do not appear in this analysis, because these subfamilies of midge are not found in those continents.) The divergences between the midges of the different continents can therefore be explained as the result of vicariance, the ocean barriers between the continents having appeared after the midges had colonized them. This in turn had useful implications as to the apparent geological ages of the different groups of midge, because the dates of separation of the continents were known from the geophysical data. (Divergence between species within each continent could, of course, have been the result of either vicariance or dispersal.) Like any method based upon an evolutionary, phyletic cladogram, Brundin's method depends entirely upon the accuracy of that cladogram, which in turn depends on the taxonomist's judgement in assessing which of each pair of divergent characters is primitive and which is derived. Where there is doubt on this, there arises the possibility of alternative phyletic cladograms and of corresponding variations in the taxon–area cladogram.

Cladistic biogeography

Some biologists have attempted to avoid the above limitations of phylogenetic cladistics by discarding any assumptions as to the evolutionary relationships of the groups under analysis. This approach is variously known as 'transformed cladism' or 'pattern cladism', and it forms the basis for cladistic biogeography, which itself includes two different approaches.

Firstly, if the phyletic and deduced biogeographic histories of several groups that occupied the area in question over the same period of time are compared, illuminating points of similarity or difference may emerge. The British palaeontologist Colin Patterson [6], for example, has shown interesting parallels between the taxon–area cladograms of several groups, suggesting that these all reflect the same sequence of separation of the areas concerned (Fig. 9.3). Secondly, one can avoid the subjectiveness of the taxonomist's judgement of evolutionary change by using as

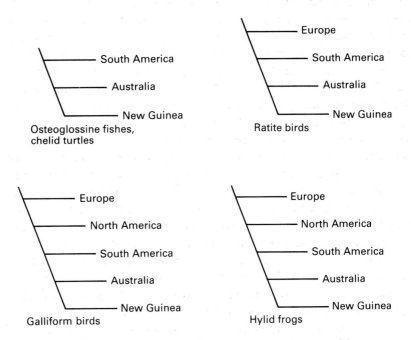

Fig. 9.3 Area cladograms of several vertebrate groups, after Patterson [6].

many different characteristics as possible, in the reasonable expectation that any anomalies in individual characteristics will become obvious and can be evaluated against the general pattern.

In applying the resulting information to biogeographical problems, data from as many groups as possible are again used in the hope that individual anomalies will become statistically unimportant. This method, and the variety of methods that can be used to evaluate and integrate the data, are well described by Humphries *et al.* [7]. These authors also give an interesting and satisfying example of the application of this method to the biogeographic history of 25 species of the eucalypt tree *Monocalyptus* along the coastal region of Southern Australia, correlating this with the climatic and geological history of the area (Fig. 9.4). It is hypothesized that the genus was distributed continuously through this area in the Early Cenozoic (uppermost figure), and that climatic change caused the progressive reduction in its range until there are now 25 different species distributed in the separate areas A–H (lowermost figure).

Panbiogeography

The alternative approach to historical biogeography commences with the analysis and comparison of patterns of distribution of organisms rather than with their relationships (cladistic or otherwise). It originated with the work of the Venezualan botanist Leon Croizat in the middle of the 1950s. Like many biologists of his time, working before acceptance of the theory of plate tectonics, Croizat was puzzled by the widely disjunct distributions of many taxa, especially in the Pacific and Indian Oceans, though he did not concern himself with the detailed phylogenetic relationships within the groups. He amassed a vast array of distributional data, representing each biogeographic pattern as a line connecting its known areas of distribution. Where these lines were similar, he combined them into what he termed a 'generalized track'. He logically deduced that such similarities in biogeographical relationships of unrelated taxa suggested that all those concerned had been affected by the appearance of the same barriers, which had interrupted a once-continuous pattern of distribution. These generalized tracks also converged on particular areas that Croizat called 'gates', 'major nodes' or 'centres of dispersal', and which he interpreted as regions where more than biogeographic system had converged [8]. He realized that these patterns of disjunct distribution could not be explained by dispersal, and he suggested an underlying geological mechanism. But, even after plate tectonics became well documented and widely accepted, Croizat for a long time refused to accept that theory, preferring other geological explanations.

Today, most biogeographers agree that plate tectonics has played a

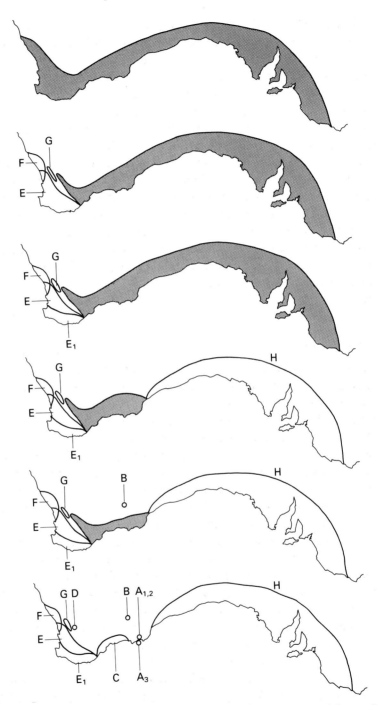

Fig. 9.4 The hypothetical sequence of changes in distribution pattern of the eucalypt tree *Monocalyptus* in southern Australia, after Humphries *et al.* [7]. The area marked in black is the continuous, hypothetical range of one particular species at each period of time from the earliest (top) to today (bottom). This becomes progressively, vicariantly subdivided into the ranges of the living species, A–H.

major role in producing the disjunct distributions that Croizat so assidu-
ously catalogued. Similarly, most of his 'nodes' are areas such as the
Panama Isthmus, South-East Asia or New Guinea, where plate tectonic
movements have brought together two areas of land of quite different
biogeographic histories, with a consequent interaction and exchange of
biota. So Croizat's ideas are not now generally accepted, but they served
a useful function in provoking the debate that led to the emergence of
the whole vicariance school of historical biogeography.

Croizat also concentrated primarily upon the distribution patterns of
living organisms, and paid little attention to the implications of the
fossil record or of climatic change. His 'generalized tracks', linking the
present-day areas within which the taxa are distributed, therefore often
cross ocean basins. For example, a generalized track runs directly across
the North Pacific between northern North America and Asia, rather
than through the intervening land regions of Siberia, Alaska and northern
Canada. This is because most forms of life are absent from these high
northern latitudes — though that is only a comparatively recent phenomenon, having been caused by the recent Ice Ages. This aspect of Croizat's
approach has more recently been emphasized by a group of mainly New
Zealand biogeographers such as R. Craw [9], who specifically require
that the track should be a 'minimum spanning graph or tree' in which
the different localities are connected by the shortest possible series of
straight lines. These tracks cross or run around the edges of the Indian
Ocean or Pacific Ocean basins and are therefore regarded as examples of
a set of distribution patterns related to those ocean basins. These sets
are referred to as 'baselines' (Fig. 9.5). This focus upon the oceans that
border the land masses through which the terrestrial organisms must
normally have spread, rather than upon the land masses themselves,
does not seem appropriate.

This panbiogeographic method also considers the area where a taxon
is most diverse in numbers, genotypes or morphology as the area from
which the track for that particular taxon radiated — a dangerous assumption (see p. 250). For example, because six species of the bird *Aegotheles*
are found in New Guinea, but only one in Australia, Craw [9] concludes
that the related fossil genus *Megaegotheles* of New Zealand was the
result of a dispersal there from New Guinea rather than from Australia
(Fig. 9.6).

Phyletic tracks

In rejecting both an evolutionary basis and a plate tectonic basis of
analysis, and in taking little account of the fossil record or of climatic
change, panbiogeographers seem to have moved so far from the methods
of other historical biogeographers that the two groups seem to be of

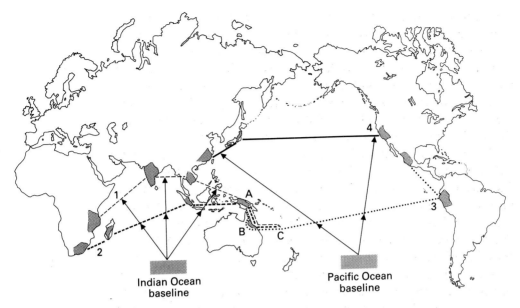

Fig. 9.5 Craw's panbiogeographic method. Tracks 1 and 2 circumscribe or cross the Indian Ocean, and represent examples of an Indian Ocean baseline, while tracks 3 and 4 similarly represent examples of a Pacific Ocean baseline. From Craw [9].

little relevance to one another. Tracks based on monophyletic groups (that is, in which all the members are descended from a single ancestral taxon) and that take past geological and climatic history into account can, in contrast, provide useful insights — though a persistent problem for any taxonomist or cladistic biogeographer is that comparatively few groups have yet been subjected to the detailed analysis that is required to establish monophyly. The American zoologist Donn Rosen has used this technique to identify a series of generalized tracks involving the Caribbean region, which imply the relationships of its terrestrial biota to North America, South America and West Africa, and of its marine fauna to the east Pacific and west Atlantic [10]. He then attempts to integrate these data with contemporary theories of the geological evolution of the Caribbean and of the Panama Isthmus to provide a vicariance model for the historical biogeography of the region over the last 150 million years (Fig. 9.7).

Endemicity and history

A completely different approach that, like panbiogeography, is not based upon phylogenetic analysis, takes as its starting-point the pattern of endemicity of a group of related taxa. The British palaeontologist Brian Rosen [11] used the method of parsimony (see above) to set up a cladistic

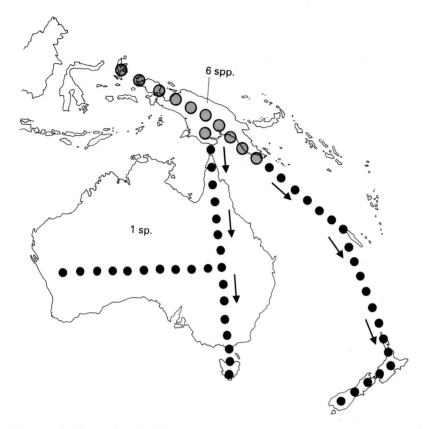

6 spp.

1 sp.

Fig. 9.6 The dispersal track of the bird genus *Aegotheles* in New Guinea and Australia, and of its fossil relative *Megaegotheles* in New Zealand. After Craw [9].

tree of the similarities between different localities, two localities being more similar if they share a greater number of taxa, and less similar if they share fewer taxa (Fig. 9.8). A closer similarity between two localities may result either from greater ecological similarity or from a more recent biotic linkage. Choice between these two possibilities as the preferred explanation depends upon the extent to which the groups concerned share a limited range of ecological requirements (suggesting an ecological explanation), or range widely across a variety of environments (suggesting a historical explanation).

The method has been used by Alan Myers of University College, Cork, to analyse the biogeographic affinities of the Hawaiian amphipods (a type of marine crustacean), some of which live there in the reef coral while others live in marine algae [12]. He found an interesting change of pattern over time, in which the biogeographic relationship of the Hawaiian species is closest to those of the south-west Pacific, in which there are many islands that could act as a source for the Hawaiian taxa. But the affinities of the Hawaiian genera instead lay with the eastern

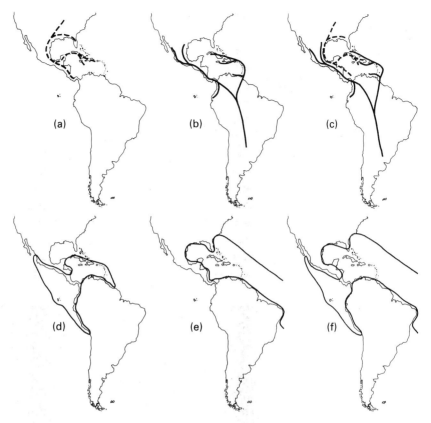

Fig. 9.7 Summary of the generalized tracks representing different elements in the historical biogeography of the Caribbean biota. (a) North American–Caribbean track. (b) South American–Caribbean track. (c) Overlapping of (a) and (b), to enclose the Caribbean Sea. (d) Eastern Pacific–Caribbean track. (e) Western Atlantic–eastern Atlantic track. (f) Eastern Pacific–eastern Atlantic track. From Rosen [10].

Pacific and the Caribbean. On average, different genera are likely to have diverged from one another earlier than their different species. It would therefore appear that the affinities of the Hawaiian genera reflect an earlier period of time, before the Panama Isthmus had closed during the Pliocene, and when there was free connection between the East Pacific and the Caribbean.

Pleistocene problems

The interpretation of the relationship between modern biogeography and that of the Pleistocene provides unique problems. The amount of data on Pleistocene distributions is quite considerable, and much greater than that on more distant geological periods. The organisms themselves are usually closely similar to those of today, so that their ecological

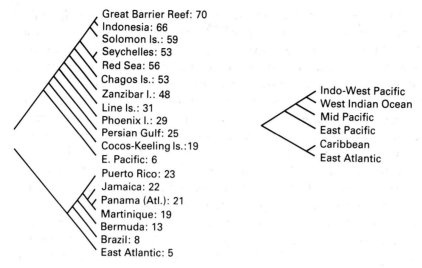

Fig. 9.8 Left, locality cladogram of living reef corals. Numbers after the locality names indicate the number of taxa that this locality shares with at least one other locality. Right, simplified cladogram, showing the general area relationships that emerge from the detailed cladogram. After Rosen [11].

preferences and modes of dispersal are well understood. Furthermore, there was little significant geographical change during and after the Pleistocene, so that it can be assumed that the present-day connections between continents were in existence — even if their potential for individual groups was different, owing to climatic alterations. But those climatic alterations were severe, affected most of the world's biota, and caused large-scale alterations in their patterns of distribution over a comparatively short space of time. As a result, the scale of the problems of interpretation that arise is more similar to those of historical palaeobiogeography, though our knowledge of the organisms themselves is more like that experienced in ecological biogeographical problems.

The identification of regions within which a number of taxa are endemic (see above) also provides a possible method of reconstructing the biogeography of areas in the recent past, during the Pleistocene glaciations. Organisms whose ecological preferences were adversely affected by these climatic changes are likely to have found their once-continuous patterns of distribution changed. They would, instead, be restricted to smaller and probably scattered areas that still preserved the climatic conditions that these organisms required. The phenomenon has long been recognized in the Northern Hemisphere, where such organisms are known as glacial relics (see p. 39).

More recently, this scenario has been proposed as a potential explanation for the unparalleled biotic diversity of the tropical rain forests of the Amazon Basin. In 1969 the American zoologist Jürgen Haffer suggested that the biogeography of many Amazon forest birds showed two features [13]. Firstly, there are a few (about six) areas that each contain clusters of endemic species with rather restricted and similar ranges: these centres of endemism together contain about 150 species of bird, which make up 25 per cent of the forest bird fauna of the Amazon Basin. Secondly, Haffer also found evidence, between these centres, of zones where the related species of the centres of endemism hybridized. He hypothesized that, during the glacial periods of the Northern Hemisphere, the centres of endemicity had been islands of persistently high rainfall (and therefore of rain forest), surrounded by areas of grassland (Fig. 9.9.) Many bird species had been able to survive in each of these refugia, where they had evolved towards becoming separate species in each. But they had not yet become fully separate species before the climatic improvement had allowed the forest and its fauna to spread again over the intervening grasslands. As a result, when the related forms met again, they were still able to hybridize.

The patterns of endemism and hybridization in Amazon birds noted by Haffer have also been found by other workers on Amazon frogs, lizards, butterflies and flowering plants. However, as the British ecologist Paul Colinvaux [14] has pointed out, though not very much is known as to the palaeoecological history of the Amazon Basin, what little is known contradicts Haffer's belief that some areas had remained with unchanged climate and vegetation to provide refuges in which some rain forest species could survive. On the contrary, the Amazon Basin seems to have undergone considerable and fundamental ecological change during the Holocene.

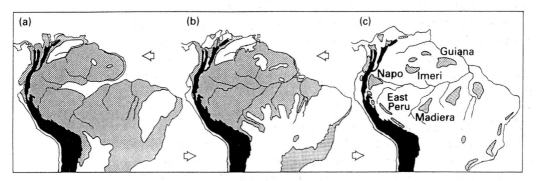

Fig. 9.9 Maps of northern South America, the shaded areas representing the lowland forests during (a) the glacial minima, (b) today and (c) glacial maxima. Coastlines have been adjusted to reflect the changes in ocean levels. Black areas represent land above 1000 m. The pattern of Amazon Basin drainage is left unaltered in all three maps, to facilitate comparison. After Lynch [15].

Centres of dispersal

At one time some biogeographers believed that the area in which a group was represented by the largest number of species was also likely to be the area from which the group dispersed. This hypothesis, however, assumes that new species will appear at a constant rate, whatever the environmental conditions, and that the presence of a large number of species in a particular area therefore indicates that the group has existed for a long time in that area. In fact the rate of speciation depends upon ecological opportunity as well as upon the appearance of new features by genetic change. A group may first appear in an area where opportunities for its particular way of life are limited, and few species will evolve. Later, it may gain access to an area in which the opportunities are far greater, and will there undergo rapid speciation. For example, though cichlid fishes are more varied in the African Great Lakes than anywhere else (see p. 151), no biogeographer today would argue that these lakes were the centres from which these fish dispersed over Africa and South America.

Some biogeographers have used the term 'dispersal centre' in a different way. By plotting the breeding ranges of many species, belonging to different groups, Paul Müller found that these overlap in one or a few smaller areas [16]. He suggests that these were centres, to which these species were confined during the Ice Ages and from which they later dispersed as the climate improved. However, Colinvaux's criticisms of Haffer's theory (above) largely apply to Müller's theory also.

Fossils and historical biogeography

In the case of living organisms, comparisons between the biota of different localities can be based upon firm data as to the definitions of the localities, the presence or absence of the taxa concerned, and upon the ecology and dispersal potential of the taxa. The biota itself often includes a wide range of organisms, so diminishing the impact on the data of the occasional aberrant dispersal phenomenon or incorrect taxonomic assignment. The organisms themselves also provide a wealth of data that can be used to construct cladistic phylogenies or patterns of endemicity. The resulting biogeographical data can then be used to suggest past patterns of geographical relationship, or to provide support for particular geographical patterns that have been suggested on geological grounds. The existence of a fossil relative of one of these groups may then provide further support for such a theory of historical biogeography, or indicate that the theory requires modification. But these fossils alone will not have been used in the initial formulation of the theory. This is because the density of documentation of the pattern of fossil distribution is usually not sufficiently detailed to provide significant data on the past geographical relationships of localities within a single modern continent. At the other extreme, at the larger scale of intercontinental relationships, there have been few changes in these geographical relationships over the last 50 million years, during which the dominant terrestrial groups have achieved their current patterns of distribution, so that there have been few incompatibilities between geography and biogeography. (The Great American Interchange provides an interesting exception to this, for the total dissimilarity between the mammal faunas of North and South America over most of the 65 million years of the Cenozoic Era had long convinced biogeographers that the Panama Isthmus connection between the two continents was of comparatively recent origin: see pp. 221–223.)

Palaeobiogeography

The interpretation of the historical biogeography, or 'palaeobiogeography', of the distant ancestors of living groups or of ancient, totally extinct groups is more difficult, bacause the data are more limited. It is often spread over a great period of geological time, during which the earth's geography was radically different from that of today, and involves groups whose ecology, powers of dispersal and phylogenetic relationships are imperfectly known. These past patterns, such as the Permo-Carboniferous patterns of plant geography (see p. 181) were historically important in provoking the realization that continents might have changed their positions and interrelationships in the past. More recently, palaeontological studies have similarly led to the realization that some areas down the western flank of North America are fragments of originally more westerly-placed, isolated packages of Pacific sea floor or islands, that later collided with North America. These areas are known as 'displaced terranes' [17]. More frequently, however, analysis of the historical biogeography of fossil groups depends on a theory of historical palaeogeography to provide the units in time and space between which biotic comparisons may provide support for the geologists' theories. A combination of knowledge of plate tectonic movements and of the shallow epicontinental seas that form additional barriers to the dispersal of terrestrial organisms can identify land areas within which useful analyses of palaeobiogeography can be made. For example, the existence of two separate Late Cretaceous Northern Hemisphere land masses, 'Asiamerica' and 'Euramerica', is supported both by geological evidence and by evidence from the distribution of contemporary dinosaurs, early mammals and plant spores, while the earliest, Permo-Carboniferous land vertebrates seem to have been limited to an earlier version of the Euramerican continent (see pp. 181–189).

References

1 Myers, A.A. & Giller, P.S. (eds) (1988) *Analytical Biogeography*. Chapman & Hall, London.

2 Rosen, B.R. (1988) Biogeographic patterns: a perceptual overview. In *Analytical Biogeography* (eds A.A. Myers & P.S. Giller). Chapman & Hall, London, pp. 23−55.

3 McKenna, M.C. (1973) Sweepstakes, filters, corridors, Noah's ark and beached Viking funeral ships in paleogeography. In *Implications of Continental Drift to the Earth Sciences*, Vol. I (eds D.H. Tarling & S.K. Runcorn). Academic Press, London & New York, pp. 295−308.

4 McKenna, M.C. (1975) Fossil mammals and Early Eocene North Atlantic land continuity. *Annals of the Missouri Botanic Garden* **62**,335−353.

5 Brundin, L.Z. (1988) Phylogenetic biogeography. In *Analytical Biogeography* (eds A.A. Myers & P.S. Giller). Chapman & Hall, London, pp. 343−369.

6 Patterson, C. (1981) Methods of paleobiogeography. In *Vicariance Biogeography: a Critique* (eds G. Nelson & D.E. Rosen). Columbia University Press, New York, pp. 446−489.

7 Humphries, C.J., Ladiges, P.Y., Roos, M. & Zandee, M. (1988) Cladistic biogeography. In *Analytical Biogeography* (eds A.A. Myers & P.S. Giller). Chapman & Hall, London, pp. 371−404.

8 Croizat, L., Nelson, G. & Rosen, D.E. (1974) Centers of origin and related concepts. *Systematic Zoology* **23**,265−287.

9 Craw, R. (1988) Panbiogeography: method and synthesis in biogeography. In *Analytical Biogeography* (eds A.A. Myers & P.S. Giller). Chapman & Hall, London, pp. 405−435.

10 Rosen, D.E. (1975) A vicariance model of Caribbean biogeography. *Systematic Zoology* **24**,431−464.

11 Rosen, B.R. (1988) From fossils to earth history: applied historical biogeography. In *Analytical biogeography* (eds A.A. Myers & P.S. Giller). Chapman & Hall, London, pp. 437−481.

12 Myers, A.A. (1990) How did Hawaii accumulate its biota? A test from the Amphipoda. *Global Ecology and Biogeography Letters* **1**,24−29.

13 Haffer, J. (1969) Specification on Amazonian forest birds. *Science* **165**,131−137.

14 Colinvaux, P. (1987) Amazon diversity in the light of the paleoecological record. *Quarternary Science Reviews* **6**,93−114.

15 Lynch, J.D. (1988) Refugia. In *Analytical Biogeography* (eds A.A. Myers & P.S. Giller). Chapman & Hall, London, pp. 311−342.

16 Müller, P. (1973) *The Dispersal Centres of Terrestrial Vertebrates in the Neotropical Region*. Junk, The Hague.

17 Cox, C.B. (1990) New geological theories and old biogeographical problems. *Journal of Biogeography*, **17**,117−130.

10 *Ice and change*

Many landform features in the temperate areas of the world show that major, geologically rapid changes in climate have taken place since the Pliocene. The general cooling of world climate that started early in the Tertiary continued into the Quaternary; the boundary between the two is placed at about two million years ago, but difficulties in definition as well as in dating techniques and geological correlation leave this date open to some doubt. The definition of the boundary comes from Italian marine sediments, where the appearance of fossils of cold-water organisms (certain foraminifera and molluscs) suggests a fairly sudden cooling of the climate. Similar evidence of cooling has been found in sediments from the Netherlands, and this is believed to mark the end of the final stage of the Pliocene (locally termed the Reuverian) and the first stage of the Pleistocene (the Pretiglian) [1]. Evidence from sediment cores in the North Atlantic indicates that débris was being carried into deep water by ice rafts as long ago as 2·5 million years. Certainly by 2·37 million years ago, the continents bordering the North Atlantic must have been experiencing periodic glaciation [2].

At various stages during the Pleistocene, ice covered Canada and parts of the United States, northern Europe, and Asia. In addition, independent centres of glaciation were formed in low-latitude mountains, such as the Alps, Himalayas and Andes, and in New Zealand. A number of present-day geological features show the effects of such glaciations; one of the most conspicuous of these is the *glacial drift deposit, boulder clay,* or *till* covering large areas and sometimes extending to great depths. This is usually a clay material containing quantities of rounded and scarred boulders and pebbles, and geologists consider it to be the detritus deposited during the melting and retreat of a glacier. The most important feature of this till, and the one by which it may be distinguished from other geological deposits, is that its constituents are completely mixed — the finest clay and small pebbles are found together with large boulders. Often the rocks found in such deposits originated many hundreds of miles away, and were carried there by the slow-moving glaciers. Fossils are rare, but occasional sandy pockets have been found that contain mollusc shells of an Arctic type. Some enclosed bands of peat or freshwater sediments within these tills provide evidence of the warmer intervals. They often show that there were phases of locally increased plant productivity, and they may contain fossils indicative of warmer climates.

Many of the valleys of a hilly glaciated area have a distinctive, smoothly rounded profile, because they were scoured into that shape by the abrasive pressure of the moving ice. In places the ice movement has left deep scratches upon the rocks over which it has passed, and tributary valleys may end abruptly, high up a main valley side, because the ice has removed the lower ends of the tributary valleys. Such landscape features provide the geomorphologist with evidence of past glaciation.

Immediately outside the areas of glaciation were regions which experienced *periglacial* conditions. These were very cold and their soils were constantly disturbed by the action of frost. When water freezes in the soil it expands, raising the surface of the ground into a series of domes and ridges. Stones within the soil lose heat rapidly when the temperature falls, and the water freezing around them has the effect of forcing them to the surface, where they often become arranged in stone stripes and polygons. Similar patterns are produced by ice wedges that form in ground subjected to very low temperatures. Sometimes these patterns, which are so evident in present-day areas of periglacial climate (see Fig. 10.1), can be found in parts of the world that are now much warmer. For example, they have been discovered in eastern parts of Britain as a result of air photographic survey. Such 'fossil' periglacial features show that, as the glaciers expanded, so the periglacial zones were pushed before them towards the Equator.

Climatic wiggles

The Pleistocene Epoch, however, has not been one long cold spell. The

Fig. 10.1 Polygons in Arctic Canada.

careful examination of tills and the orientation of stones embedded within them soon showed that several advances of ice have taken place during the Pleistocene, often moving in different directions. Occasional layers of organic material were sometimes discovered trapped between tills and other deposits, and these have provided fossil evidence of warm periods alternating with the cold. Where sequences of deposits are reasonably complete and undisturbed, as in the eastern part of England (East Anglia) and parts of the Netherlands, it has been possible to construct schemes to describe these alternations of warm and cold episodes, to name them and to determine their relationships in time. But in many parts of the world this has not proved at all easy, and the correlation of events between different areas has often been speculative and unsatisfactory, mainly because of the difficulty experienced in obtaining secure dates for the deposits. At one time, for example, geologists considered that there were four episodes of ice advance in Europe, defined mainly by sequences of tills in the Alps. The four glaciations were named Günz, Mindel, Riss and Würm. This is now regarded as a simplification of the true situation, and it seems likely that there have been far more climatic fluctuations in the Pleistocene than this simple model suggests.

Because of the difficulties experienced in climatic reconstruction using land-based (terrestrial) evidence, attention has turned to the seas, where marine sediments provide a more complete and uninterrupted sequence. The retrieval of long, deep ocean cores of sediment has provided an opportunity to follow the rise and fall of various members of plankton communities in the past, particularly those, like the foraminifera which, although tiny, have robust outer cases that survive the long process of sedimentation to the ocean floor and there accumulate as fossil assemblages. Some members of the foraminifera, like some species of *Globigerina* and *Globorotalia*, are sensitive to ocean temperature, so their relative abundance in the fossil record provides evidence of past climates.

An even more powerful tool for reconstructing long-term climatic changes has been the use of oxygen isotopes retained in the fossil material of the sediments. 'Normal' oxygen (^{16}O) is far more abundant than the heavier form of oxygen (^{18}O). For example, the heavy form comprises about 0·2 per cent of the oxygen incorporated into the structure of water (H_2O). Water evaporates from the sea, but those molecules containing ^{18}O condense from a vapour form rather more readily than their lighter counterparts, so this heavy form tends to return rapidly to the oceans. Water containing ^{16}O, on the other hand, remains in the atmosphere as vapour for longer and is more likely to fall eventually over the ice caps and to become incorporated into these as ice. Under cold conditions, the volume of global ice increases and this (since it is formed largely from precipitation) tends to lock up more of the ^{16}O, leaving the oceans richer in ^{18}O. So the ratio of $^{18}O : ^{16}O$ left in the oceans increases during periods of cold. This ratio is then reflected in the skeletons of foraminifera and other planktonic organisms and is deposited in the ocean beds. Analysis of the oxygen isotope ratios in ocean sediments thus provides a long and continuous record of changing water temperatures going back millions of years. It has even been possible to use these methods for the analysis of oxygen isotope ratios in inland areas, as in the gradual deposition of calcite in the Devil's Hole fault in Nevada [3].

As a result of such oxygen isotope studies of a series of cores in the Caribbean and Atlantic Oceans, Cesare Emiliani of the University of Miami, Florida, has been able to contruct palaeotemperature curves for the ocean surface waters [4]. A summary curve for the past 700 000 years is shown in Figure 10.2 and it is quite obvious from this diagram that the climatic changes in this latter part of the Pleistocene have been numerous and complex. Figure 10.2 also shows a long sequence derived from the analysis of a deep-sea core from the equatorial Pacific Ocean by Nick Shackleton of the University of Cambridge, England. This analysis covers the past two million years, and the upper part can be correlated with the Emiliani curve. Such correlation ('wiggle matching') is assisted by the magnetic reversals that have occurred during the course of the Quaternary and which provide a basic time framework. A magnetic reversal is a situation where the polarity of the earth's magnetic field is switched, leaving a record in the rocks because of those particles incorporated in the sediments that had become magnetized by the earth's field. These reversals have been occurring throughout the earth's history, and five have taken place in the last million years, the latest one about 700 000 years ago. They provide very useful datum horizons for the correlation of sediments.

Examination of the long core in Fig. 10.2 (a) shows that the fluctuations in temperature become stronger as one advances through the Quaternary. Gentle wanderings around a mean value gradually become more pronounced as greater extremes of temperature are experienced. Sharper changes also become more apparent, with the temperature likely to change both radically and rapidly. One can also see that the course of change, although showing a general wave-like form, is not simple or regular, but has many minor wiggle patterns imposed upon it. This latter point implies that we should not expect the terrestrial record in currently temperate lands such as North America or central Europe to show a simple alternation of temperate conditions with arctic ones, but a much more varied pattern in which cold and warmth alternate in varying degrees and in which intermediate conditions are often met.

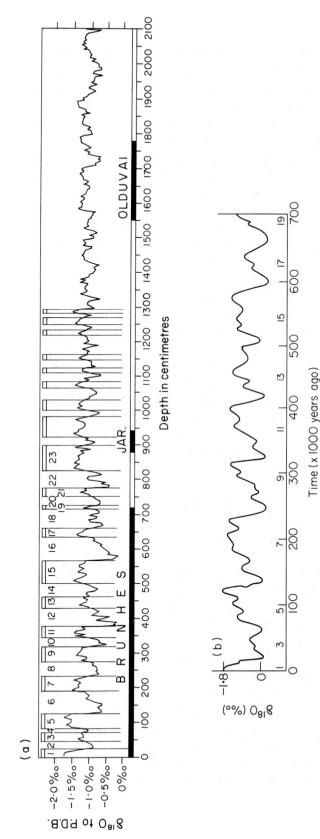

Fig. 10.2 (a) Oxygen isotope curve derived from sediments in the Pacific Ocean, which record over two million years of climatic history Lower (more negative) ^{18}O : ^{16}O ratios indicate higher temperatures. Brunhes, Jar (Jaramillo) and Olduvai black bars indicate periods of reversal of the earth's magnetic field. (b) A more detailed record from another Pacific core, covering only the last 700 000 years. After Emiliani [4]. Data from N.J. Shackleton.

Interglacials and interstadials

The fluctuations in global temperature represented in the ocean cores are reflected in the terrestrial geological sequence by glacial and inter-glacial deposits. A warm episode (usually represented by an organic, peaty deposit), sandwiched between two glacial events (often represented by tills) and which achieved sufficient warmth for a sufficient duration for temperate vegetation to establish itself, is termed an *interglacial*. The sequence of events demonstrated in the fossil material of such an interglacial shows a progressive change from high arctic conditions (virtually no life), through sub-arctic (tundra vegetation) to boreal (birch and pine forest) to temperate (deciduous forest) and then back through boreal to arctic conditions once more. If the warm event is of only short duration, or if the temperatures attained are not sufficiently high, then the vegetation changes may only reach a boreal stage of development. In this case it is termed an *interstadial*. We are currently living in the most recent interglacial (termed by geologists the Holocene).

Interglacials are often times of increased biological productivity (except in the more arid parts of the world) and they are often represented in temperate geological sequences as bands of organic material (see Fig. 10.3). This material usually contains the fossil remains of the plants, animals and microbes that existed at or near the site during its formation, and it is this evidence, often stratified in a time sequence, that allows us to reconstruct past conditions and habitats. One of the most valuable sources of fossil evidence for this purpose has been the pollen grains from plants that have been preserved in the sediment and that reflect the vegetation of the area at that time. Pollen grains are sculptured in such a way that they are often recognizable with a con-siderable degree of precision; they are also produced in large numbers (especially those of wind-pollinated plants) and are widely dispersed; finally, they are preserved very effectively in waterlogged sediments, such as peats and lake deposits. So the analysis of pollen grain assem-blages can provide much information about vegetation and hence about climate and other environmental factors [5].

Fossil pollen data are usually presented in the form of a pollen diagram, and Fig. 10.4 shows a pollen diagram belonging to the last (Ipswichian) interglacial in Britain. The vertical axis is the depth of deposit, which is inversely related to time, so the diagram should be read from the bottom up. The sequence begins with boreal trees, birch and pine, which are then replaced by deciduous trees such as elm, oak, alder, maple and hazel. Later in the sequence these trees go into decline, to be replaced by hornbeam, and then pine and birch once more. The details of such a sequence will obviously vary from one locality to another (Fig. 10.5 shows the pollen diagram from the same interglacial

Fig. 10.3 Organic sediments from an interglacial (dark band), resting on gravel of a former raised beach and covered by deposits laid down under periglacial conditions. A cliff exposure at West Angle, Dyfed, Wales.

further east in Europe, in Poland, where lime, spruce and fir play a more important role) and varies quite considerably between different interglacials, but there is a consistent pattern to all such sequences, for they all pass through a predictable series of developmental stages. Often these are shown on such diagrams by dividing them into four zones (usually labelled in Roman numerals I–IV), pre-temperate, early-temperate, late-temperate and post-temperate respectively. Pollen diagrams from our present interglacial suggest that we are well advanced into the late-temperate stage.

The last interglacial is the most easy to identify stratigraphically because it occurred in the very recent past. Placing precise dates upon it is difficult because the radiocarbon method becomes less accurate as one proceeds back in time, and dating methods relying on argon isotopes are dependent on the presence of volcanic material. But it is believed that the last interglacial began about 130 000 years ago (though a recently studied and well-dated site from Nevada suggests commencement at

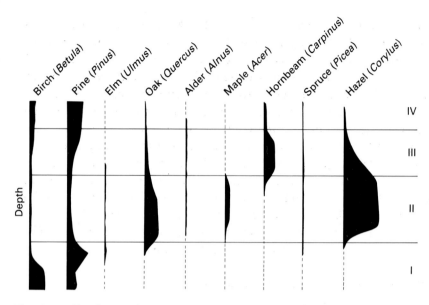

Fig. 10.4 Pollen diagram from sediments of the last (Ipswichian) interglacial in Britain. Only tree taxa are shown. The depth axis is related to age.

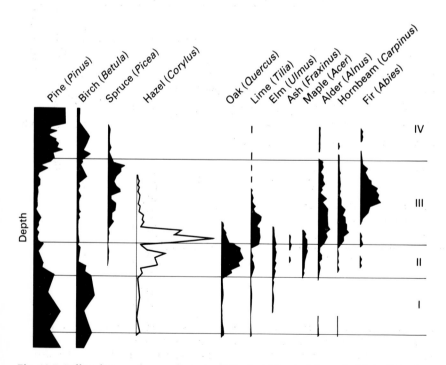

Fig. 10.5 Pollen diagram from sediments of the last (Eemian) interglacial in Poland.

147 000 years ago) and ended some 115 000 years ago with the commencement of the last glaciation. This means that we can identify the terrestrial record of the last interglacial with oxygen isotope stage 5 (Fig. 10.2(b)) in the marine sediments. Earlier interglacials are much more problematic; they are usually given local names when they are first described, and attempts are later made to correlate them, often on the basis of the fossils they contain. Some general correlations are shown in Fig. 10.6 and a more detailed attempt to correlate the glacial and interglacial sequence of the Netherlands and East Anglia (Britain) is shown in Fig. 10.7. As can be seen from the second of these schemes, the problem of correlation is made even more difficult by gaps in the record in some localities. Any attempt at a global synthesis of terrestrial data is premature without a system for supplying absolute dates. The estimated dates on Fig. 10.6 are based on magnetic reversals.

Biological changes in the Pleistocene

With the expansion of the ice sheets in high latitudes, the global pattern of vegetation was considerably disturbed. Many areas now occupied by temperate deciduous forests were either glaciated or bore tundra vegetation [3]. For example, most of the north European plain probably

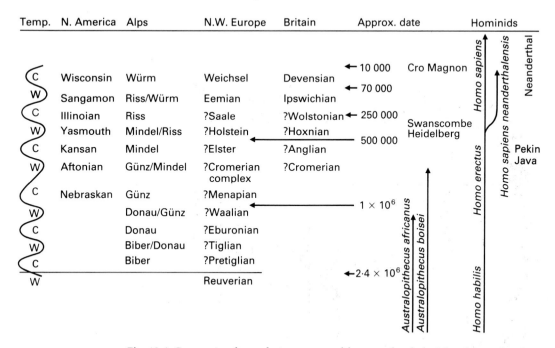

Fig. 10.6 Conventional correlations assumed between local glacial and interglacial stages in the Pleistocene. Local complexities render such correlations tentative, particularly in the earlier stages. See Fig. 10.7 for a suggested correlation of European stages in the early part of the Pleistocene. C, cold; W, warm.

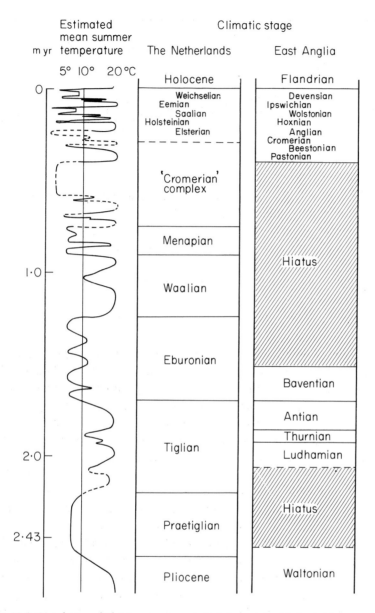

Fig. 10.7 Correlation of Pleistocene stages in Britain and the Netherlands as proposed by Zagwijn. A suggested temperature curve is also shown. After West [1].

had no deciduous oak forest during the glacial advances. The situation in Europe was made more complex by the additional centres of glaciation in the Alps and the Pyrenees. These would have resulted in the isolation and often the ultimate local extinction of species and, indeed, of whole communities of warmth-demanding plants during the glacial peaks. Figure 10.8 shows the broad vegetation types that occupied Europe

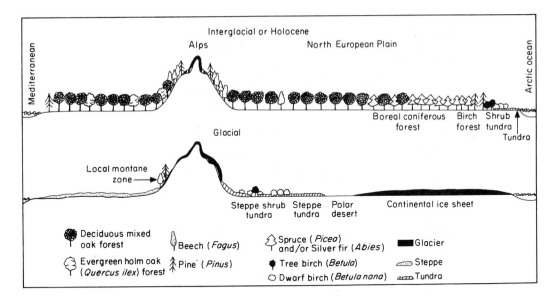

Fig. 10.8 The vegetation belts of Europe during glacial and interglacial time.

during interglacial and glacial times. During the interglacials, tundra species would have become restricted in distribution due to their inability to cope with such climatic problems as high summer temperatures or drought, and their failure to compete with more robust, productive species. High-altitude sites and disturbed areas would have served as *refugia* within which groups of such species may have survived in isolated localities. Similarly, during glacials particularly favourable sites which were sheltered, south-facing, or oceanic and relatively frost-free may have acted as refugia for warmth-demanding species [6].

Extinctions did occur, however. In Europe many of the warmth-preferring species so abundant in the Tertiary Epoch were lost to the flora. The hemlock (*Tsuga*) and the tulip tree (*Liriodendron*) were lost in this way, but both survived in North America, where the generally north–south orientation of the major mountain chains (the Rockies and the Appalachians) allowed the southward migration of sensitive species during the glacials and their survival in what is now Central America. The east–west orientation of the Alps and Pyrenees in Europe permitted no such easy escape to the south. The wing nut tree (*Pterocarya*) was also extinguished in Europe, but it has survived in Asia, in the Caucasus, and in Iran.

The last glacial

The most recent glacial stage lasted from approximately 115 000 years ago until about 10 000 years ago, so our current experience of a warm earth is really rather an unusual episode as far as recent geological history is concerned. But even a glacial is not a period of uniform cold, and there have been numerous warmer interruptions to the prevailing cold climate. Many interstadials from the temperate areas of the world are recorded within the last glacial.

In the tropics the effect of ice was not, of course, experienced directly except on the very high mountains, but climate was generally colder, and changes in vegetation reflected in pollen diagrams suggest that there were important variations in precipitation. In Queensland, Australia, for example, Peter Kershaw has analysed the sediments of a volcanic crater lake from a rain forest area, and the results are shown in Fig. 10.9 (pages 268–9). Dating at this site is difficult, but the total span of the diagram is thought to cover about 120 000 years, extending back into the closing stages of the last interglacial. At that time many of the types of rain forest tree that currently occupy the area were growing but, at the time when the final glacial began in high latitudes, this forest was replaced by forest of simpler structure in which *Cordyline* plays an important part. There is then a brief reversion (dated at about 86 000–79 000 years ago) when rain forest became re-established for a spell, but then much more arid conditions set in and the forest became increasingly dominated by gymnosperm trees, such as the monkey puzzle (*Araucaria*). Between about 26 000 and 10 000 years ago the climate became very dry and the former forests took on a sclerophyllous form, being dominated by *Casuarina*, a tree currently associated with hot dry conditions. At the end of the 'glacial', however, there was a very rapid change in vegetation, with the invasion of rain forest trees once again.

The period of maximum aridity at this site, 26 000–10 000 years ago, encompasses the period of maximum extent of the Northern Hemisphere glaciers at about 20 000 years ago. Tropical aridity seems to have been widespread at this time, as is confirmed by evidence from tropical Africa, India and South America. Much of the humid tropical forest of the Zaire basin was probably replaced by dry grassland and savanna during the glacial, although some forest may well have survived in riverine and lakeside situations. The temperature records from tropical locations, such as East Africa, follow closely the trends observed in other parts of the globe.

Figure 10.10 shows the approximate areas of rain forest that existed at the time of maximum glacial extent in the high latitudes (about 20 000 years ago). From this it can be seen that rain forest was much fragmented as a result of the glacial drought in tropical latitudes [8]. Many areas currently occupied by rain forest were reduced to savanna woodlands at this time, and this is an important point to bear in mind when considering the high species diversity of the rain forests (see

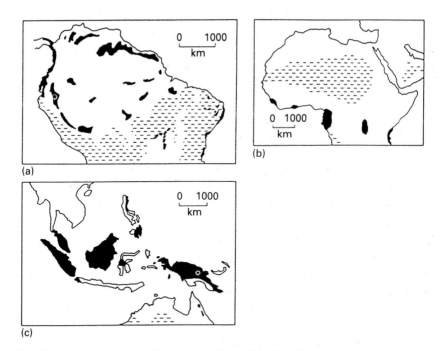

Fig. 10.10 Suggested areas of humid rain forest (black) and arid/semi-arid areas (hatched) 20 000 years ago, when the glaciation of the higher latitudes was at its maximum. From Tallis [8].

Chapter 2). Most rain forests have not enjoyed a long and uninterrupted history but have been disrupted by global climatic changes. But whereas the temperate forests were forced to occupy new areas at lower latitudes, the tropical forests had nowhere to which they could retreat, so became fragmented and dismembered. Some ecologists believe that this fragmentation may even have added to their diversity by permitting the isolation of populations and the development of new species.

Figure 10.10 also shows areas of more extreme drought at the period of maximum glacial extent, and the global pattern of dry regions is shown in greater detail in Fig. 10.11, where it is compared to modern desert distribution [9]. The global aridity of the last stages of the last glacial is very apparent. Blown sand (*loess*) from these deserts is found fossil in many parts of the eastern United States and in central Europe. Lake-level studies have similarly shown that the period between 15 000 and 20 000 years ago was particularly dry. For example, the lake levels from tropical Africa were at a low point during this time period, as shown by the data summarized in Fig. 10.12.

It would be misleading, however, to give the impression that drought prevailed throughout the earth at this time. Just as at the present day some areas are wetter than others, so it was in glacial times, and the

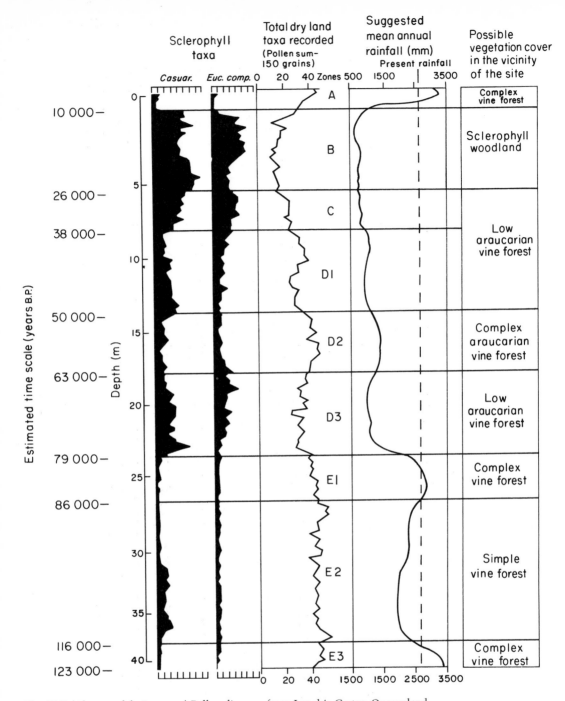

Fig. 10.9 (Above and facing page.) Pollen diagram from Lynch's Crater, Queensland, Australia. The column on the left shows the percentage of rain-forest gymnosperms (stippled); rain-forest angiosperms (white); sclerophyll taxa (hatched). The frequencies of pollen of all taxa are shown as percentages of the dry land plant pollen total; each division represents 10 per cent of the pollen sum. Abbreviations for taxa: *Arau.*, *Araucaria*; *Da.*, *Dacrydium*; *Podoc.*, *Podocarpus*; *Cordyl.*, *Cordyline comp.*; *Cu.*, *Cunoniaceae*; *El.*, *Elaeocarpus*; *Fr.*, *Freycinettia*; *Rapan.*, *Rapanea*; *Casuar.*, *Casuarina*; *Euc. comp.*, *Eucalyptus comp.* After Kershaw [7].

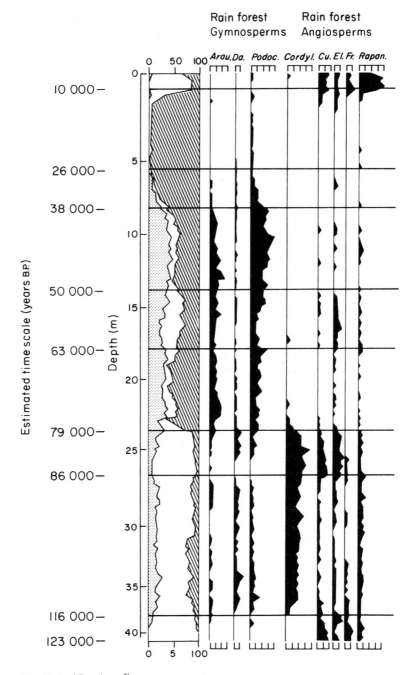

Fig. 10.9 (Continued)

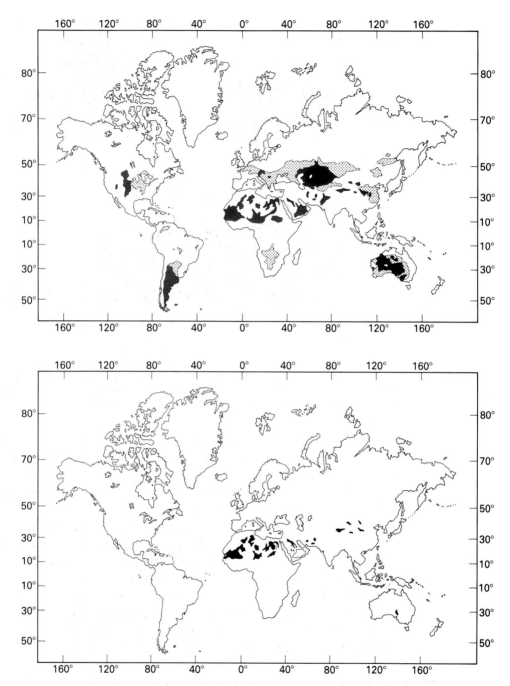

Fig. 10.11 Distribution of sand deserts at the present day (below) and suggested distribution at the height of the last Ice Age (above). Mobile sand (loess) was a feature of many areas during the glacial episode. From Wells [9].

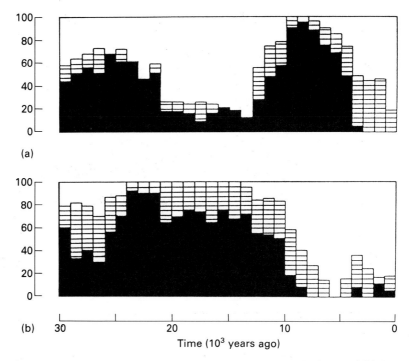

Fig. 10.12 Lake levels over the past 30 000 years in (a) tropical Africa and (b) the western United States. The bars represent the proportion of lakes studied with high (black), intermediate (shaded) or low (white) lake levels. From Tallis [8].

western part of the United States enjoyed a time of wet climate and high lake levels between 10 000 and 25 000 years ago. Such wet periods are called 'pluvials', and many parts of the world experienced pluvials at various times in Quaternary history. In Africa, for example, 55 000 and 90 000 years ago seem to have been times of pluvial climate.

Evidence for the existence of pluvial lakes in western North America is provided not only by the geological deposits but also by the present-day distribution of certain freshwater animals. In western Nevada there are many large lake basins that are now nearly dry, but in the remaining waterholes there live species of the desert pupfish (*Cyprinodon*) (see Fig. 10.13). Over 20 populations of the pupfish are known in an area of about 3000 square miles [10]. The isolated populations have gradually evolved into what are considered four different species, each adapted to its own specific environment, rather like the Hawaiian honey-creepers (see p. 161). In many respects the wet sites in which these fish live can be regarded as evolutionary 'islands' separated from each other by unfavourable terrain. The species have probably been isolated from one another since the last pluvial at the beginning of the present interglacial, whereas populations within each species may still be in partial contact during periods of flooding. The pluvial periods of the Pleistocene

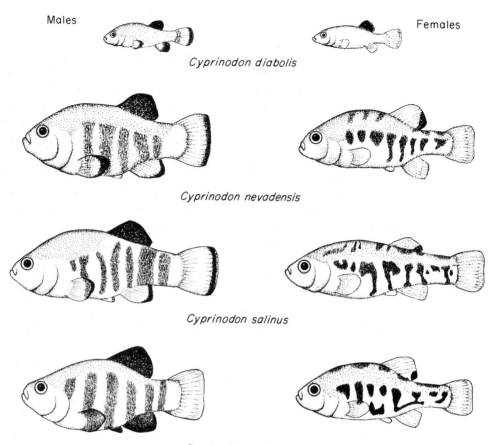

Males   Females

Cyprinodon diabolis

Cyprinodon nevadensis

Cyprinodon salinus

Cyprinodon radiosus

Fig. 10.13 The four species of desert pupfish that have evolved in the streams and thermal springs of the Death Valley region. Males are bright iridescent blue; females are greenish. After Brown [10].

provided the conditions necessary for these aquatic animals. An Old World example is provided by the waterbug genus *Corixa*, which is now widespread in Europe, but in Africa exists in only a few scattered localities extending south to the Rift Valley in Kenya. Presumably it was much more abundant when what is now the Sahara was dotted with lakes, and has since become restricted in distribution as the climate has grown drier.

Fossil evidence from various parts of the world dating from the last glacial thus shows that the glacial stages represented times of severe disruption to the entire biosphere. This fact, coupled with the assurance that we are still locked into the oscillating climatic system that has operated over the past two million years, makes it imperative that we should understand the mechanisms that have generated the glacial/ interglacial cycle.

Causes of glaciation

Ice ages are relatively rare events in the earth's 4·5 billion years of history. Although the polar regions receive less energy from the sun than the equatorial regions, they have been supplied with warmth by a free circulation of ocean currents through most of the world's history. Only occasionally do land masses pass over the poles, or form obstructions to the movement of waters into the high latitudes. The movement of Antarctica into its position over the South Pole led to the development of a southern polar ice cap perhaps as long ago as 15 million years. Rearrangements of the Northern Hemisphere land masses have subsequently led to the isolation of the Arctic Ocean so that it received relatively little influence from warm-water currents, leading to its becoming frozen some three to five million years ago. The presence of two such polar ice caps increased the albedo, or reflectivity, of the earth, for whereas the earth as a whole reflects about 40 per cent of the energy that falls on it, the ice caps reflect about 80 per cent. The formation of two such ice caps therefore significantly reduced the amount of energy retained by the earth. The scene was set for the development of an 'Ice Age'.

It is still necessary, however, to explain why the Ice Age has not been a period of uniform cold, but has consisted of a sequence of alternating warm and cold episodes. Perhaps the most acceptable proposal to explain this pattern is that put forward in the 1930s by the Yugoslav physicist Milutin Milankovich, though the idea had been in circulation and had even been supported by Alfred Wegener of continental drift fame. Milankovich constructed a model based on the fact that the earth's orbit around the sun is elliptical and that the shape of the ellipse changes in space in a regular fashion, from more circular to strongly elliptical. When the orbit is fairly circular there will be a more regular input of energy to the earth through the year, whereas when it is strongly elliptical the contrast between winter and summer energy supply will be much more pronounced. It takes about 100 000 years to complete a cycle of this change in orbital shape.

A second source of variation is produced by the tilt of the earth's axis relative to the sun, which again affects the impact of seasonal changes, with a cycle duration of about 40 000 years. The third consideration is a wobble of the earth's axis around its basic tilt angle, which shows a cycle of about 21 000 years. The pattern of climatic change should, according to Milankovich, be a predictable consequence of summing the effects of these three cycles.

Geophysicists working on the chemistry of ocean floor sediments have expended much effort in seeking evidence for cyclic periodicity in past ocean temperatures and in checking the cycles found against those proposed by Milankovich. In 1976 Jim Hays, John Imbrie and Nick Shackleton [11] were able to confirm that all three levels of Milankovich cycle, 100 000, 40 000 and 21 000 years, could be detected in the sediments.

The Milankovich pattern has now been found in sediments dating back eight million years, but one important change has been detected in their effects. Whereas eight million years ago the effects of the 100 000-year cycle were weak, in the last two million years it is this cycle that has been very strong and that has dominated the glacial/interglacial sequence. There must be additional factors that have amplified this particular cycle in recent times [12].

One possible explanation for the current exaggeration of the effects of the 100 000-year cycle is that ice masses themselves are responsible. The ice grows slowly and decays relatively quickly, which can itself modify the global climate. Computer models have been constructed which take account of this effect, and they produce a better fit to observed data than the Milankovich cycles on their own [13].

Detailed testing of the Milankovitch theory is, of course, dependent on well-dated records and at present the dating of climatic fluctuations in the Pleistocene is crude. The difference in the dating of the commencement of the last interglacial, for example, varies over 17 000 years depending on the materials used. Until dating is more firmly established, full confirmation of the correlation between orbital cycles and climate cannot be achieved.

A further complicating factor in the elucidation of the causes of climatic change is that a shift in temperature can induce changes in the global carbon cycle that have feedback effects. The importance of this process has been revealed by direct measurements of the past atmosphere by the analysis of gas bubbles trapped in the Greenland ice cap [14]. Figure 10.14 shows the results of the analysis of carbon dioxide (top curve) and methane (bottom curve) in these gas bubbles, and the curves are compared with the proposed temperature curve (middle) for the last 160 000 years. It is very evident that both methane and carbon dioxide were elevated during the last interglacial to levels very similar to those of the present interglacial, and that they were depressed during the glacial episode. Even the minor fluctuations in these gases correspond quite closely to the proposed variations in temperature during the last glacial.

It is not surprising that an increase in warmth should be accompanied by increases in the atmospheric load of these two gases, for both are associated with the activities of organisms. Carbon dioxide is produced by respiratory activity of plants, animals and microbes, and one might well anticipate elevated levels during interglacials. The ocean also acts as an important reservoir for carbon, and raised carbon dioxide levels in

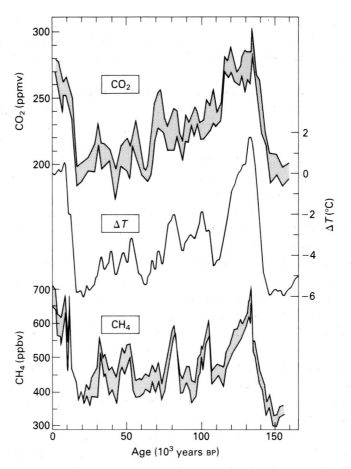

Fig. 10.14 Analyses of gas bubbles in an ice core taken from the Antarctic ice cap. These are regarded as fossil samples of the ambient atmosphere of the past. The concentrations of carbon dioxide (upper) and methane (lower) are shown together with a reconstruction of temperature change derived from oxygen isotope studies of the same area (middle). From Lorius *et al.* [14].

the atmosphere may be related to a different behaviour of the oceanic mixing processes in glacials and interglacials. Methane is generated by partial decomposition in wetlands and is also produced by ruminant animals and by termites in significant quantities, so it too might be expected to increase in warm periods. But both gases are strongly absorptive of infra-red radiation, so can contribute strongly to a 'greenhouse effect' in which they insulate the earth and enhance its energy retention. The outcome of this process is that warm periods generate methane and carbon dioxide, and the presence of these gases in the atmosphere further elevates global temperature, so that there is a positive feedback and the temperature fluctuation is further amplified. Some climatologists believe that the rising levels of methane and carbon dioxide slightly precede the development of a warm episode and actually drive the climatic variations.

A possible consequence of these fluctuations in atmospheric carbon dioxide is that the anatomy of some plants may have altered in response to the changes. Since plants take up the gas through the pores (stomata) in their leaves, they may have needed fewer stomata under conditions of high carbon dioxide concentration. Ian Woodward of Sheffield University in Britain has shown that plants have changed their stomatal density in parallel with CO_2 changes in the last century [15], and the search is now on for evidence of such changes in the more distant past. Such work should at least provide further evidence for the atmospheric changes projected from ice core analysis.

Wallace S. Broecker of Columbia University has emphasized the importance of the circulation of the earth's ocean currents, coupled with those of the atmosphere, to explain the rapidity with which some of the climatic shifts in the Pleistocene have come about [16]. His theory does not contradict the ideas of Milankovitch, but supplements them. Much of the warmth of the North Atlantic is brought into the region by highly saline waters heading northwards at intermediate depths of about 800 metres. These waters are brought to the surface in the neighbourhood of Iceland, mainly as a result of wind action, and their exposure to the cold air causes them to give up their heat and then, being cold, saline and dense, they descend to the ocean floor. This deep, saline water then makes its way southwards, rounds the southern tip of Africa, and spreads through the Indian and Pacific Oceans (Fig. 10.15). As it warms it rises, and its direction is reversed so that it returns to the North Atlantic. Broecker estimates that this input of energy into the North Atlantic is equivalent to 30 per cent of the annual solar energy input to the area. Fossil evidence, however, indicates that this oceanic 'conveyor belt' became switched off during the glacial episodes, and the reduction in the heat transfer to high latitudes led to the development of the extended ice sheets. How this switch is effected is still not fully understood, but one possibility is an alteration in the salinity and therefore the density of ocean waters, and an example of this will be described in Chapter 11.

A further suggestion is that massive volcanic eruptions may precede and initiate the process of glaciation [17]. There is certainly some evidence for a correlation between glacial advances and periods of volcanic activity during the last 42 000 years in New Zealand, Japan and South America. Volcanic eruptions produce large quantities of dust, which are thrown high into the atmosphere. This has the effect of reducing the amount of solar energy arriving at the earth's surface, and dust particles also serve as nuclei on which condensation of water droplets occurs, thus increasing precipitation. Both of these consequences

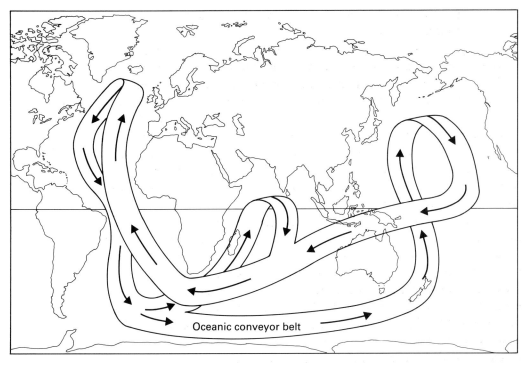

Fig. 10.15 Scheme showing the circulation of the oceans. The system is driven by changes in temperature and salinity of the waters. As warm, low-salinity waters move northwards up the Atlantic, they become cold and dense and return as a deep-water current. This passes along southern latitudes into the Pacific and returns as warmer, surface water.

would favour glacial development. Attempts to correlate volcanic ash content with evidence of climatic changes in ocean sediments, however, have not met with much success [18], except to show an increase in the general frequency of volcanic activity during the last two million years.

References

1 West, R.G. (1977) *Pleistocene Geology and Biology*, 2nd edn. Longman, London.
2 Shackleton, N.J. *et al.* (1984) Oxygen isotope calibration of the onset of ice-rafting and history of glaciation in the North Atlantic region. *Nature* **307**, 620–623.
3 Winograd, I.J., Szabo, B.J., Coplen, T.B. & Riggs, A.C. (1988) A 250 000-year climatic record from Great Basin vein calcite: implications for Milankovitch theory. *Science* **242**, 1275–1280.
4 Emiliani, C. (1972) Quaternary paleotemperatures and the duration of high-temperature intervals. *Science* **178**, 398–401.
5 Moore, P.D., Webb, J.A. & Collinson, M.E. (1991) *Pollen Analysis*, 2nd edn. Blackwell Scientific Publications, Oxford.
6 Delcourt, H.R. & Delcourt, P.A. (1991) *Quarternary Ecology: A Paleoecological Perspective*. Chapman & Hall, London.

7 Kershaw, A.P. (1974) A long continuous pollen sequence from north-eastern Australia. *Nature* **251**,222–223.

8 Tallis, J.H. (1991) *Plant Community History*. Chapman & Hall, London.

9 Wells, G. (1989) Observing earth's environment from space. In *The Fragile Environment* (eds L. Friday & R. Laskey). Cambridge University Press, Cambridge,

10 Brown, J.H. (1971) The desert pupfish. *Scientific American* **225**(11),104–110.

11 Hays, J.D., Imbrie, J. & Shackleton, N.J. (1976) Variations in the earth's orbit: pacemaker of the ice ages. *Science* **194**,1121–1132.

12 Kerr, R.A. (1981) Milankovitch climate cycles: old and unsteady. *Science* **213**,1095–1096.

13 Denton, G.H. & Hughes, T.J. (1983) Milankovitch theory of ice ages: hypothesis of ice-sheet linkage between regional insolation and global climate. *Quaternary Research* **20**,125–144.

14 Lorius, C., Jouzel, J., Raynaud, D., Hansen, J. & Le Treut, H. (1990) The ice-core record: climate sensitivity and future greenhouse warming. *Nature* **347**,139–145.

15 Woodward, F.I. (1987) Stomatal numbers are sensitive to increases in CO_2 from pre-industrial levels. *Nature* **327**,617–618.

16 Broecker, W.S. & Denton, G.H. (1990) What drives glacial cycles? *Scientific American* **262**(1),43–51.

17 Bray, J.R. (1976) Volcanic triggering of glaciation. *Nature* **260**,414–415.

18 Ninkovitch, D. (1976) Explosive Cenozoic volcanism and climatic implications. *Science* **194**,899–906.

11 *From the present into the future*

Following its maximum extent at about 20 000 years ago, the climate began to warm and the great ice sheets started to recede. Every indication pointed to the commencement of a new interglacial. Our present interglacial has been a very brief episode in the history of the earth, yet it has seen changes over the face of the planet that have taken place at a faster rate than ever previously observed. The last 10 000 years or so differ from the rest of the earth's history in one important respect; they have seen the rise to a position of ecological dominance of a single, influential species, *Homo sapiens*. During this interglacial our species has emerged from being one among a number of socially organized predators into a position from which it can and does modify whole global cyles and climates. It is no exaggeration to state that there is no habitat or ecological process on earth that can now be studied without reference to the impact of this species. It is natural, therefore, that this short, final part of the earth's history demands a biogeographer's special attention.

The current interglacial — a false start

Before one can undertake a survey of the rise of our species, it is worth pausing to consider the climatic instability that typifies the closing stages of the last glaciation and the opening of the current interglacial, for its tells us something of the rapidity with which climate can change, and may serve as a warning about future changes. This instability was first noted by geologists in Denmark, who found that the sediments of lakes dating from the transition between the glacial and the current interglacial exhibited some unusual features. The inorganic clays, typical of sediments formed in lakes surrounded by arctic tundra vegetation where soils are easily eroded and the vegetation has a low organic productivity, were replaced by increasingly organic sediments as the development of local vegetation stabilized the mineral soils, and the aquatic productivity of the lakes led to an increasing organic content in the sediments. But this process, reflecting the increased warmth of the climate, was evidently interrupted, for the sediments then reverted to heavy clay, often with angular fragments of rock denoting a return to severe climatic conditions. Ultimately, the organic sediments reappeared and the climatic warming became evident once again, this time leading to the development of our present interglacial, but the interruption proved to be a consistent feature of sediments of this age throughout north-west Europe.

The cold episode that caused the deposition of these sediments was severe enough to cause the regrowth of many glaciers, and geomorphologists have shown that the glaciers of Scandinavia and Scotland extended considerably during this episode, while small glaciers began to form on north-facing slopes in more southerly mountains, from Wales to the Pyrenees. The event was called the 'Younger Dryas', because the fossil leaves of the arctic plant the mountain avens (*Dryas octopetala*) were found in abundance in the clay layers during the original Danish studies (Fig. 11.1). Radiocarbon dating of the Younger Dryas from a range of sites indicates that it took place somewhere between 11 000 and 10 000 years ago, interrupting a warming process that had begun around 14 000 to 13 000 years ago. In a number of sites in continental Europe a shorter and less severe interruption to the warming process (called the 'Older Dryas') was also detected and dated to around 12 000 to 11 800 years

Fig. 11.1 Mountain avens (*Dryas octopetala*), an arctic—alpine plant, remains of which were found in sediments from Denmark and which has lent its name to the Younger Dryas cold event.

ago, but its impact was geographically more confined and its influence was much weaker than that of the Younger Dryas.

Although stratigraphic and fossil evidence for the Younger Dryas cold phase was abundant in north-west Europe, it was more difficult to discern in the sediments of the Alps, especially on the southern side. Sediments from the Pyrenees show a distinct Younger Dryas, but on the other side of the Atlantic its influence is difficult to detect when one moves inland from eastern North America. The evidence, therefore, suggests that the event was centred upon the North Atlantic, and oceanic evidence from fossils in the sediments shows that a cold front of polar water did indeed extend southwards to Iberia during this time.

An elegant explanation of how this climatic reversion, centred in the North Atlantic, may have come about was put forward by Claes Rooth of the University of Miami and was later supported by the data of Wallace Broecker [1] and his co-workers. They suggest that the change was brought about by a change in the pattern of discharge of meltwater from the great Laurentide glacier of North America. As this ice mass, covering the whole of eastern Canada, melted, its waters flowed initially down to the Gulf of Mexico. During the Younger Dryas, however, it is proposed that the meltwater was rerouted via the St Lawrence into the North Atlantic (Fig. 11.2). Not only would this bring large volumes of cold water into the North Atlantic, but the fresh water would also dilute the saline waters on the oceanic conveyor belt (see p. 274) and could disrupt the global movement of this conveyor. Indeed, modifications to the conveyor could prove an important component in many climatic changes [2]. Such modification would explain why the Younger Dryas is most strongly felt in the regions around the North Atlantic, where the warm influence of the conveyor has its greatest impact.

If the conveyor were disrupted in this way, however, one might expect its effects to be felt globally. Within the last few years there have been reports from as far afield as Alaska [3] and the South China Sea [4] which do document a cooling during Younger Dryas time, so this is also consistent with the Rooth/Broecker model. But not all the evidence is quite so supportive. Richard Fairbanks [5] of Columbia University has conducted some very detailed work on the corals of Barbados, and from them has been able to reconstruct past sea levels with great precision. At the height of the last glaciation, sea levels around Barbados were over 120 metres lower than today because so much of the earth's water was incorporated into ice caps and glaciers. From the rate of sea level rise during deglaciation, one can calculate the rate of meltwater flow into the oceans, and Fairbank's data indicate that the greatest rate of discharge was at about 12 000 years ago (while the Mississippi carried it to the Gulf of Mexico) and had declined by 11 000 years ago (the com-

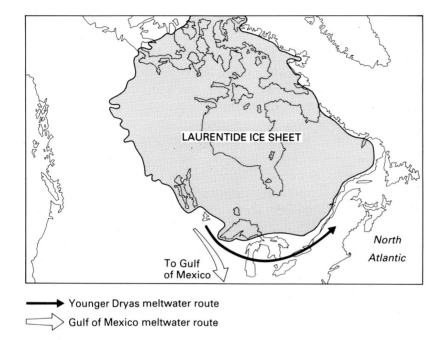

Fig. 11.2 The melting Laurentide ice sheet in the final stages of the last Ice Age, showing alternative routes for the discharge of meltwater. From Broeker *et al.* [1].

mencement of the Younger Dryas). It is difficult to reconcile this with the Rooth/Broecker model. The complications caused by atmospheric circulation patterns and the level of carbon dioxide in the atmosphere may well now be invoked to improve the model.

Perhaps the most interesting and important aspect of the Younger Dryas is that it demonstrates how quickly the earth's climate can flip from interglacial to glacial mode and back again. The brief warm spell before the Younger Dryas was sufficiently hot to bring beetles with Mediterranean affinities as far north as the middle of England. Yet within a few centuries the area was back in glacial conditions. Perhaps even more remarkable is the change at the end of the Younger Dryas. The work of Willie Dansgaard [6] of the University of Copenhagen, and his colleagues, on the Greenland ice cap have shown that the Younger Dryas came to an abrupt end 10 720 years ago, and that it was followed by a warming of about 7°C in only 50 years. So the detailed study of these times could help us to understand how plants and animals react to very rapid and very considerable changes in climate.

Forests on the move

After this faltering start to the current interglacial, the general rise in temperature (recorded in the oxygen isotope ratios of the ocean sediments) was consistently maintained, and the vegetation and animal life of the earth once again had to adjust to a new set of conditions. The pollen grains preserved in the accumulating lake sediments of 10 000 years ago provide detailed records of the arrival and expansion of tree populations as species distribution patterns changed and forests were reconstituted. Figure 11.3 shows a generalized pollen diagram for southern Britain, compiled from a range of sites.

In part, the sequence of trees (birch, pine, elm plus oak, lime and alder) reflects the changing climatic tolerances of species as the climate warmed, but it is also a function of their relative speed of migration and the distances which they had to cover to reinvade land exposed by the retreating glaciers. Birch, for example, has light airborne fruits which can travel considerable distances. It is also able to produce fruits when only a few years old, which permits a rapid expansion of its range. Add to this the fact that it is cold-tolerant and may have survived on the mainland of Europe quite close to the British Isles during the last glaciation, and it is not surprising that birch should be the first tree to appear in any abundance in the postglacial pollen record. One must also remember that birch produces large quantities of pollen and that one

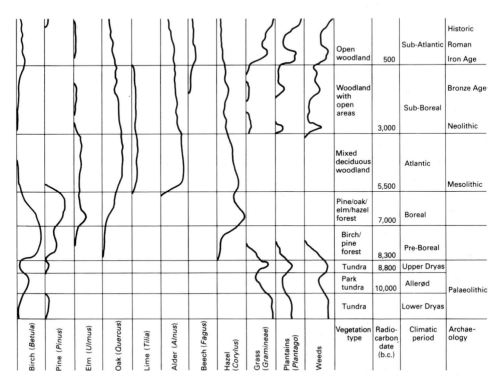

Fig. 11.3 Generalized pollen diagram from the southern part of the British Isles, showing the changing proportions of tree pollen and some non-arboreal types during the final stages of the Pleistocene and through the Holocene.

species, *Betula pubescens*, is likely to have been growing locally around the lake sites in which pollen-bearing sediments are found. Many factors, therefore, must be considered before a pollen sequence such as that displayed in Fig. 11.3 can be interpreted in terms of changing climate. Some elegant techniques have recently been employed to try to translate pollen densities in sediments into estimates of the population densities. In this way one can trace the population expansion of trees as they invaded new areas [7].

Overall, it is evident from the data that there was a warming of the climate during the early stages of the interglacial, reaching a maximum between 7000 and 5000 years ago (5000–3000 BC). This warmth maximum is marked by the invasion of lime (*Tilia*), which extended its range as far as southern Scotland and locally even further north at this time.

Large numbers of pollen diagrams covering the current interglacial are now available from all over the world and many of them are firmly dated by means of radiocarbon. This has made it possible to study the movement of individual species and genera of plants by constructing pollen maps for particular periods in the past. In this way one can follow the spread of trees, for example, and observe the routes along which they have dispersed and the way in which they have reassembled themselves into reconstituted forests. Figure 11.4 shows some examples of this type of study taken from the work of George Jacobson, Tom Webb and Eric Grimm on the recolonization of America by trees after the last glaciation [8].

Just three tree taxa have been selected from their extensive studies to illustrate the patterns of tree movement in the last 18000 years. Spruce survived the glacial maximum in the mid-west and along a broad front immediately to the south of the Laurentide ice sheet. It followed close on the retreating ice front, eventually settling in Canada. Pollen levels suggest that it has increased in density over the past 6000 to 8000 years, just as it did in the later stages of many earlier interglacials. Pine is more difficult to interpret because there are several species all with very different climatic requirements, and they cannot be effectively separated on the basis of their pollen grains. Other fossil material, such as their needles, however, indicates that both southern and northern pine types survived the glaciation in the south-east of the United States, mainly on the Atlantic coastal plain. The two groups subsequently separated, one group heading north to invade areas left bare by retreating ice, and the other group becoming firmly established in Florida. The oaks had found refuge from the glaciation in Florida, where they had achieved a marked dominance by 8000 years ago. Subsequently, they have declined in Florida and have moved into their current stronghold to the south of the Great Lakes area.

The technique of mapping tree movements has provided some valu-

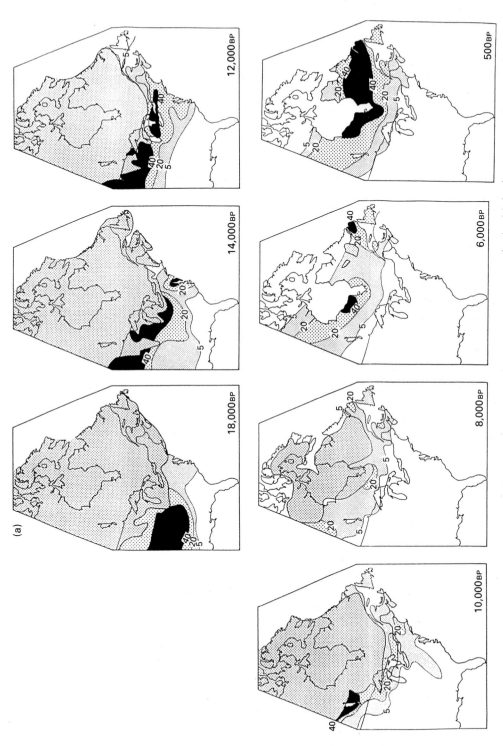

Fig. 11.4 Spread of selected trees through North America since the last glacial maximum. Dates are in years Before Present. Contours ('isopolls') join sites of equal pollen representation in lake sediments from the appropriate time period. From Jacobson *et al.* [8]. (a) Spruce (*Picea*). BP = before present.

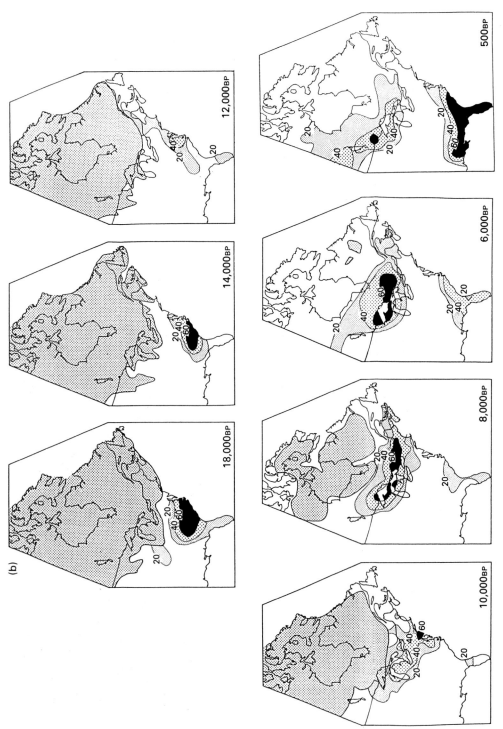

Fig. 11.4 (b) Pine (*Pinus*).

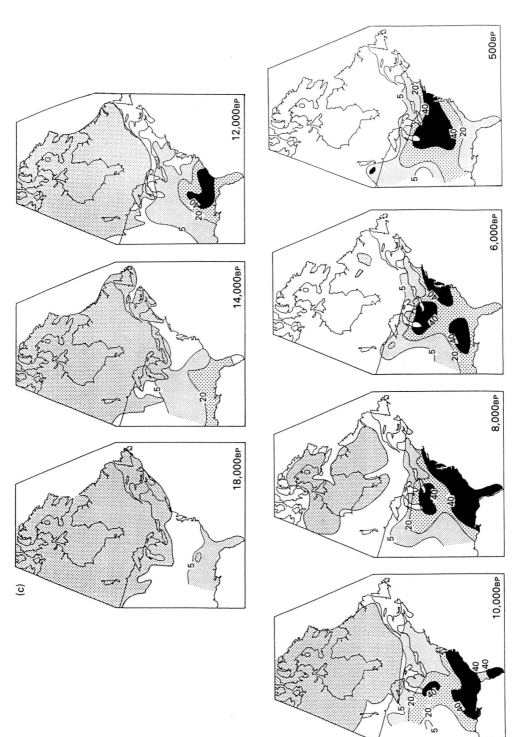

Fig. 11.4 (c) Oak (*Quercus*)

able information about the rapidity with which different species can respond to climatic change, which may well prove useful when we are concerned with predicting future responses to our currently changing climate. The rates of spread in response to climate change, even of large-seeded species such as the oaks, are surprisingly rapid. In Europe, for example, oak spread reached 500 metres per year in the early part of the current interglacial. Rates of 300 metres per year are common for many tree and shrub species.

The emergence of modern humans

It was not only trees and other plants that moved into the higher latitudes from their glacial refuges to take advantage of the new lands being released from their former ice cover. Many animal species expanded into the new regions and among them was our own species. *Homo sapiens* evolved during the late Pleistocene, and the arrival of this species on the scene proved of such dramatic ecological importance that it is necessary to interrupt the narrative of the current interglacial to trace the emergence of our species.

The fossil history of humans is very incomplete, but each year brings new fossil material which helps to fill in the gaps and provides a more detailed picture of how anatomically modern humans emerged. The primates of the New World and the Old World became separated some 40 million years ago, and it is the Old World branch that was ancestral to humans. The hominoids, which include such living primates as the orang-utan, the chimpanzee and gorilla, first began their development in Africa and spread into Asia when the two continents collided at the end of the Miocene, some 17 million years ago (see p. 224). The last common ancestor of the true hominids and our nearest living relative, the chimpanzee, lived about 8 million years ago [9].

The hominids themselves are first known in East Africa, in Tanzania and in Ethiopia, about 3·75 million years ago. Among the fossils of this age is the partial skeleton which has come to be known as 'Lucy', and which has supplied a great deal of anatomical information about the early hominids. These fossils have been assigned the name *Australopithecus afarensis*, and it is believed that they walked upright on their hind legs — a conclusion which is supported by the extraordinary discovery of fossil human footprints in volcanic ash from Tanzania [10]. The habitat in which these organisms lived was open woodland and savanna, far from the dense tropical forests, but very little is known of their precise way of life and the ecological niche which they occupied in the ecosystem. Walking on two legs, it is probable that they used tools with their free hands, much in the way that a chimpanzee is capable of doing, but there is no evidence that they were able to make extensive modifications to natural tools. They probably contented themselves with using the sticks and stones which they found.

But would they have used such tools for digging up vegetable roots, for scavenging for carrion, or for hunting living prey? As yet there is much speculation on this point, but little in the way of direct evidence. Whatever its ecological role in the ecosystem, we must regard this primitive hominid as one more species in the complex food web of the ecosystem which it occupied. There is no reason to believe that it was more influential than any other of its fellow species.

By two million years ago, evolutionary developments had taken place in the hominid line and there were at least three species of hominids present in East Africa: two of these were species of *Australopithecus*, but the third belonged to our own genus and has been called *Homo habilis*. This species had a larger brain than its relatives, and there is evidence that it was able to modify stone tools, thus showing a marked cultural change which was to continue with remarkable consequences along the *Homo* line of development. Knowledge of the diet and the ecological role of *Homo habilis* is still fragmentary, but it is believed that the meat content of the diet increased.

At 1·75 million years ago we find that *Homo habilis* has vanished and has been replaced by *Homo erectus*, and there is every reason to believe that this new species was the direct evolutionary product of *Homo habilis*. There are two aspects of *Homo erectus* that are of biogeographical significance: first its cultural developments and secondly its distributional changes. On the cultural side we find much more sophisticated stone tools, including structures which could be termed hand axes, and even more significantly there is evidence that some populations of *Homo erectus* used fire. Clearly this latter development provided the species with the capacity to modify its environment more profoundly than any other cultural development to date. Although fire would have been used as a means of food preparation, its potential as an aid in hunting must surely have been appreciated by this intelligent species, and the consequences must have been felt in terms of the new constraints placed upon the flora and fauna of its surroundings. The distribution of the species is also of interest, for this is the first species of our genus to be found outside Africa, and by one million years ago it had spread into eastern Asia, Java and southern Europe.

From this widespread species, modern humans have emerged during the last million years. There are still disputes about the precise relationship between modern *Homo sapiens* and the type known as Neanderthal [11]. It is now apparent that the two human types coexisted during the last glaciation, but only one emerged from the glacial, and that was our own species [12]. Whether this replacement actually involved the deliberate destruction of the Neanderthals, or whether there was interbreeding remains uncertain, but the fact remains that only *Homo sapiens* remained in the Holocene.

The spread of *Homo sapiens* into the New World probably took place during the last glacial episode, when sea levels were sufficiently low to expose a land bridge across what are now the Bering Straits. The exact date of arrival is still disputed, but indirect evidence from the geography of human languages suggests that the invasion must have taken place before the major advance of the last glaciation at about 20 000 years ago. The linguistic reseach of R.A. Rogers [13] has revealed three distinct groups of native American languages, and these are centred on the three ice-free refugial areas of North America during the height of the last glaciation (Fig. 11.5). It seems likely that human populations were isolated in these three areas during the glacial maximum and subsequently spread to other areas. The extinct language of Beothuk, once spoken in Newfoundland, could belong to another population isolated in that eastern refugium.

Modern humans and the megafaunal extinctions

The spread of the human species at the end of the glaciation was accompanied by an alteration in the fauna of North America and of Europe, namely the extinction of many species of large mammals — the megafauna. It was long assumed that these extinctions were the result of the climatic changes, but the American anthropologist Paul Martin suggested [14] that humans, instead, may have been the culprits. Martin pointed out that most of the animals that became extinct were large herbivorous mammals or flightless birds, weighing over 50 kg body weight — precisely the part of the fauna that humans might have been expected to hunt. Martin also pointed out that similar extinctions had taken place in other, more southern areas than North America, and suggested that the timing of these extinctions varied, in each case the time corresponding with the evolution, or arrival, of a race of humans with relatively advanced hunting techniques. In Africa, for example, where humans probably evolved, the extinctions of large herbivores apparently took place before those in the Northern Hemisphere. However, it is difficult to date these precisely, and even more difficult to correlate them with any changes in the hominid cultures of that continent. The same is true of the extinctions that took place in South America.

It is in North America that the record of extinctions has been studied in the greatest detail. Martin suggests that 35 genera of large mammal (55 species) became extinct in North America at the end of the last (Wisconsin) glaciation — over twice as many as had taken place during all the earlier glaciations, and this at a time when the climate was already improving. This certainly seemed to support the idea that

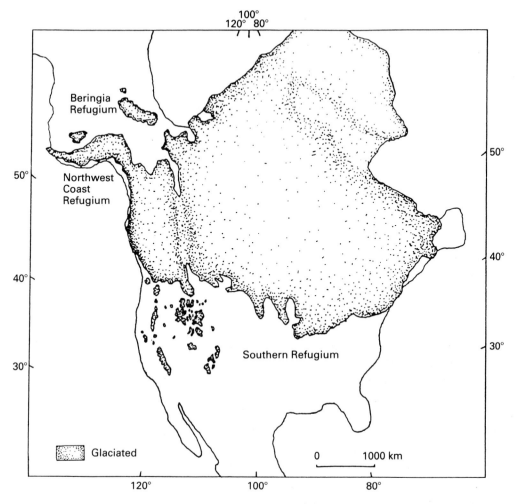

Fig. 11.5 Maximum extent of the last (Wisconsin) glacial in North America, showing the three ice-free areas (refugia) which correspond to the language groups of native Americans. From Rogers *et al.* [13].

some agent other than climate had been responsible, and it seemed reasonable to suspect the hunting activities of humans. But the American anthropologist Grayson [15] has shown that there was a similar rise in the level of extinctions of North American birds (ranging from blackbirds to eagles) at that same time. Since it is unlikely that early humans were responsible for the extinction of these birds, this observation throws doubt upon the whole hypothesis of the dominant role played by humans in Pleistocene extinctions in general.

On the other hand, the fact that so many North American species become extinct at the same time (11 500–10 500 years ago) with the arrival of hunting peoples, whereas in Europe the extinctions were spread over a longer period, provides a strong body of circumstantial evidence to support Paul Martin's claim that humans are to blame [16]. The debate continues, and the extinction of so many different species of mammal may not have been caused by one single factor, but there is an increasing number of studies in which the time of extinction can be precisely correlated with the arrival or intensification of settlement by human populations. An example is given in Fig. 11.6, showing the dates of the last recorded dung of the Shasta ground sloth from a variety of American sites [17]. Extinction seems to have taken place at many localities at about 11 000 years ago, which is when hunting cultures — Clovis Man — were most active in the region. It is difficult to escape the conclusion that the marks of human activities first became noticeable very early in the present interglacial.

Perhaps a newly developed technique which enables researchers to identify the precise source of blood on prehistoric stone weapons will shed further light on the prey animals which were most favoured by early hunting communities [18].

Domestication and agriculture

It was while these extinctions were taking place in various parts of the world that populations of humans living in south-west Asia were ex-

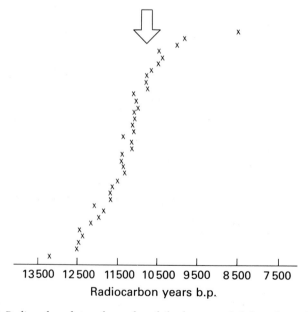

Fig. 11.6 Radiocarbon dates of samples of the last recorded dung from the Shasta ground sloth in the south-west of the United States. The arrow represents the time at which the Clovis hunters were most active in the region. From Lewin [17].

perimenting with a new technique for enhancing their food supplies. In the fertile region of Palestine and Syria grew a number of annual grasses with edible seeds — the ancestors of our wheat and barley. Like many other plants, these species were advancing north with the improving climate, and they may have proved particularly adept at exploiting the clearings that human groups would make for their camps and settlements in the developing woodland. No doubt the inhabitants of such settlements would soon have discovered the nutritional value of their seeds and have learned to propagate and ultimately cultivate them.

Tracing the ancestry of our modern species of cereals is difficult, but Fig. 11.7 represents a possible scheme for the evolution of modern

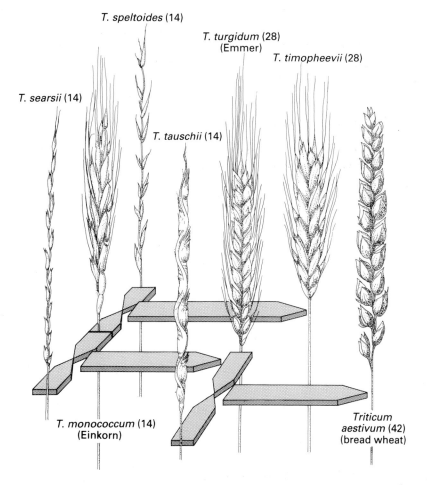

Fig. 11.7 The evolution of modern bread wheat. This reconstruction is tentative, but represents the probable course of crossings among the wild wheat species that led to the early domesticated forms of the genus *Triticum*, and the subsequent crossings of domesticated wheats with wild species and chromosome doubling that led to bread wheat. Figures in parentheses after names represent chromosome numbers.

wheat, based on a study of the chromosome numbers of the various wild and cultivated species. The original wild wheats undoubtedly had a total of 14 chromosomes (seven pairs) in each cell. *Triticum monococcum* (einkorn) was the first wild wheat to be extensively used as a crop plant. It probably hybridized with other wild species, but the hybrids would have proved infertile because the chromosomes could not pair up prior to the formation of gametes. Faulty cell division in one of these hybrids, however, could solve this problem because once the chromosome number had doubled (*polyploidy*) chromosome pairing could take place and the species would become fertile. Fertile polyploid hybrid species formed in this way included another important crop species, emmer (*T. turgidum*) with 28 chromosomes. But the bread wheat of modern times (*T. aestivum*) has 42 chromosomes and this probably arose as a result of polyploidy following the hybridization of emmer with another wild species, *Triticum tauschii*. This species comes from Iran and it probably interbred with emmer as a result of the transport of that wheat species to Iran by migrating human populations. Thus early agricultural peoples began not only to modify their environment but also to manipulate the genetics of their domestic species.

Other wild plants of the Middle East were grown and cultivated by humans (see Fig. 11.8), among them barley, rye, oat, flax, alfalfa, plum and carrot [19]. Further west, in the Mediterranean basin, yet more native plants were domesticated, including pea, lentil, bean and mangel-wurzel. Recent analysis of all the accumulated data from European archaeological investigations shows how these crop plants gradually became used more extensively across the continent and how new species were adopted in regions where climatic factors limited the use of some early types. Some of these results are shown in Fig 11.9. *Lens culinare, L. orientalis* and *L. nigricans*, the lentils, have proved useful crop species in the Near East, but climatic factors have prevented their being used extensively in northern Europe. Barley (*Hordeum vulgare*), however, proved much more suitable for cultivation in Europe than it had in the early domestication sites in south-west Asia.

The idea of plant domestication seems to have evolved independently in many different parts of the globe and at many different times. In each area, appropriate local species were exploited: in south-west Asia there were millet, soybean, radish, tea, peach, apricot, orange and lemon; central Asia had spinach, onion, garlic, almond, pear and apple; in India and South-East Asia there were rice, sugar cane, cotton and banana. Maize, New World cotton, sisal and red pepper were originally found in Mexico and the rest of Central America, while tomato, potato, common tobacco, peanut and pineapple first grew in South America.

In some cases there may have been independent cultivations of the same or similar species in different parts of the world. Thus emmer

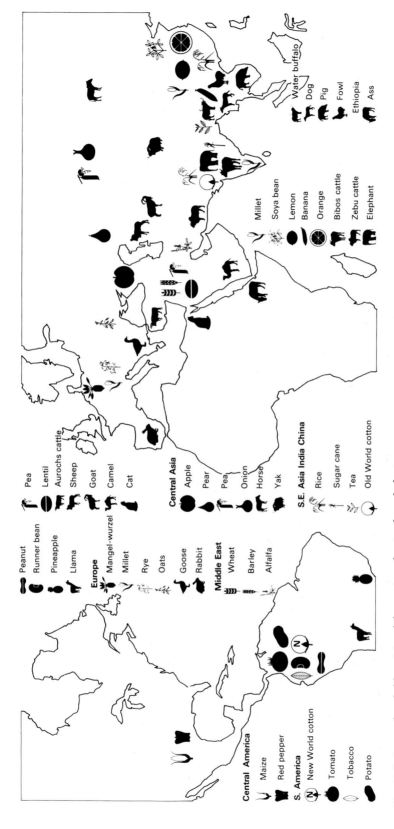

Fig. 11.8 Areas of probable origin of domesticated animals and plants.

Central America
Maize
Red pepper

S. America
New World cotton
Tomato
Tobacco
Potato

Europe
Mangel-wurzel
Millet
Rye
Oats
Goose
Rabbit

Middle East
Wheat
Barley
Alfalfa

Peanut
Runner bean
Pineapple
Llama

Pea
Lentil
Aurochs cattle
Sheep
Goat
Camel
Cat

Central Asia
Apple
Pear
Pea
Onion
Horse
Yak

S.E. Asia India China
Rice
Sugar cane
Tea
Old World cotton

Millet
Soya bean
Lemon
Banana
Orange
Bibos cattle
Zebu cattle
Elephant

Water buffalo
Dog
Pig
Fowl
Ethiopia
Ass

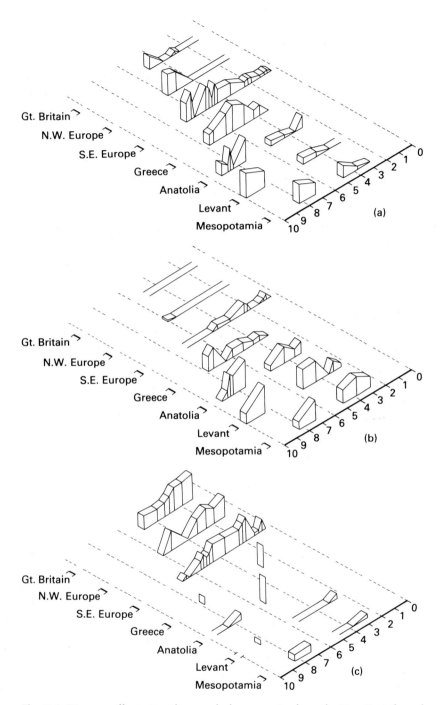

Fig. 11.9 Diagrams illustrating the spread of crop species from the Near East through Europe. The vertical scale represents the percentage of sites in which the species is present and the horizontal scale is in thousands of radiocarbon years before present. Blanks indicate a lack of data. The species represented are: (a) *Triticum monococcum,* einkorn wheat; (b) *Lens culinare, L. orientalis* and *L. nigricans,* lentils; (c) *Hordeum vulgare,* six-row hulled barley. From Hubbard [19].

wheat may well have originated quite independently in the Middle East and in Ethiopia.

One of the most puzzling problems in the study of plant domestication is the origin of maize. Unlike most of the cereals of Eurasian origin, there is no wild species that can be regarded with certainty as its progenitor. Most research workers into this subject are now agreed that the most likely ancestor is the Mexican annual grass, teosinte. Indeed, both maize and teosinte are now regarded as sub-species of *Zea mays*, but structurally they are very different, and in particular the evolution of the all-important flower and fruit structure is still in dispute. The real problem is that there are no intermediate types between teosinte and maize, and neither are there any archaeological records of wild teosinte grains being collected and used for food.

Very often maize appears quite suddenly in the diet of human populations, and it may then become the dominant food resource. This can be traced by an elegant chemical technique in which the ratio of carbon-13 to carbon-12 in fossil human bone structures is analysed. Maize is a C_4 plant and one of the characteristics of these plants is that the ratio of ^{13}C to ^{12}C in the sugars produced by photosynthesis differs from that found in C_3 plants. This difference is retained in animals feeding on the plant, so we can trace the importance of C_4 plants in an animal's diet [20].

One recent explanation of the sudden appearance of maize without any intermediate form being recorded has been put forward by the botanist Hugh Iltis [21], who claims that an abnormality arose in teosinte, possibly set off by some environmental factor, in which male inflorescences became female on the lateral branches, thus producing a kind of quantum leap in evolution.

The domestication of certain animals may have preceded that of plants. There is some evidence, for instance, that earlier cultures had domesticated the wolf or, in some North African communities, the jackal. Such animals were probably of considerable use in driving and tracking game and hunting down the wounded prey. Domesticated dog remains dating from 12 000 BC have been found in Iraq.

However, it is likely that many of the other animals that became associated with humans, such as sheep and goats, were domesticated during the early Neolithic period soon after the first cultivation of plants. These were initially herded for their meat and hides, but would have also been a source of milk, once tame enough to handle. The first traces of domesticated sheep come from Palestine around 6000 BC. These may have originated from one of the three European and Asiatic sheep, or may have resulted from interbreeding among these species. The Soay sheep of the Outer Hebrides almost certainly originated from the moufflon, either the European *Ovis musimon* or the Asiatic

O. orientalis. Domestication of these animals may have resulted from the adoption of young animals orphaned as a consequence of hunting activity. In Israel there is a marked shift in diet between 10 000 and 8000 years ago, gazelle and deer being replaced by goat and sheep. Almost certainly this was a consequence of domestication [22].

The aurochs, *Bos primigenius*, was a frequent inhabitant of the mixed deciduous woodland which was spreading north over Europe during the postglacial period. In many of the sites where remains of these forests have been preserved, such as in buried peats and submerged areas, the bones of this animal have been found. It was probably first domesticated by the Neolithic farmers of Anatolia around 7000 BC.

The dry lands

At the time of the first agricultural experiments at the opening of the Holocene, the Middle East was experiencing a time of mild, humid climate. Pollen diagrams from Syria show that oak forest was expanding into the dry, steppic vegetation that had persisted during the glacial maximum. In fact, many of those parts of the world currently occupied by desert or semi-arid scrub suffered a similarly dry climate during the glacial maximum, but the close of the glaciation brought renewed rains to these dry areas. Studies of lakes in the vicinity of the Sahara suggest that conditions became more humid in a series of stages, commencing around 14 000 years ago, but there was a short period of aridity during the Younger Dryas. The early part of the Holocene then provided a time of wetness for many currently desert areas extending from Africa through Arabia to India. Lakes existed in the middle of the very arid Rajasthan Desert of north-west India, and the surge of fresh water down the Nile created stratified waters in the eastern Mediterranean, the low-density fresh water lying over the top of the high-density salt water. As a result the lower layers became depleted in oxygen, and black anoxic sediments, called *sapropels*, were deposited [23].

The wetness permitted the northward extension of savanna and rain forest into formerly dry areas, and Fig. 11.10 summarizes data from many pollen diagrams taken from sites in the southern Sahara [24]. The expansion of the more humid biomes can be seen, followed by their contraction when aridity set in once more about 5000 years ago. The Sahara Desert is quite rich in ancient rock paintings, some dating back over 8000 years, and these older ones depict big game animals now associated with the savanna grasslands, confirming the evidence of the pollen. Between 7500 and 4500 years ago the painters of the pictures were evidently pastoralists who depicted their cattle on rocks in locations where cattle could certainly not graze today. After that time the pictures of cows were replaced by camels and horses as the arid climate became

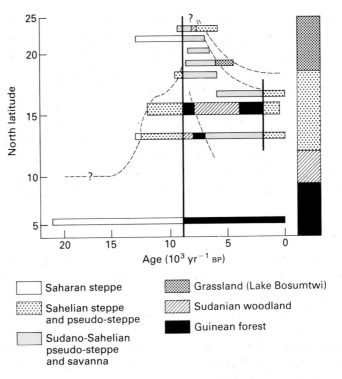

Fig. 11.10 Changes in the vegetation in northern tropical Africa over the past 20 000 years. The present latitudinal zonation of vegetation is shown on the right of the diagram and it can be seen that this zonation was displaced about 5°N during the period 9000 to 7000 years ago, in the early stages of the current interglacial. From Lezine [24].

more severe, and the habits and domesticated animals of the local peoples altered accordingly.

Many of the great deserts of the world were evidently initiated by climatic changes in the latter half of our interglacial. The involvement of increased human activity during this time, however, has obviously complicated the picture, and some researchers believe that human exploitation, even in prehistoric times, may have contributed to the development of deserts, as in the case of the Rajasthan Desert, where the increasing aridity was accompanied by the rise of the Indus Valley civilization. It can be difficult to differentiate between cause and effect in such circumstances.

Changing sea levels

The current interglacial has also been a time of changing sea levels. The melting of ice released considerable quantities of water into the oceans resulting in a (*eustatic*) rise in sea level relative to the land. This may have amounted to as much as 100 m in places. On the other hand, the loss of ice caps over those land masses which acted as centres of glaciation relieved the earth's crust of a weight burden, resulting in an (*isostatic*) upwarping of the land surface with respect to sea level. The relative importance of these processes varied from one place to another, depending upon how great a load of ice an area had borne. In western Europe the result was a general rise in the level of the southern North Sea and the English Channel with respect to the local land surface. In this way Britain, which was a peninsula of the European mainland during the glaciation, gradually became an island. Evidence from submerged peat beds in the Netherlands suggests a rapid rise in sea level between about 10 000 and 6000 years ago, which has subsequently slowed down gradually. By this latter date England's links with continental Europe would have been severed. At this time many plants with slow migration rates had still not crossed into Britain and were thus permanently excluded from the British flora. The separation of Ireland from the rest of Britain occurred rather earlier, and many species native to Britain had not established themselves as far west as Ireland. As a result, plants such as the lime tree (*Tilia cordata*), and herb paris (*Paris quadrifolia*) are not found growing wild in Ireland.

There is, however, one group of plants of great interest to plant geographers that did succeed in reaching Ireland before the rising sea level separated that country from the rest of the British Isles, and this is known as the *Lusitanian* flora. Lusitania was the name for a province of the Roman Empire consisting of Portugal and part of Spain and, as its name suggests, this flora has affinities with that of the Iberian peninsula. Some of the plants — such as the strawberry tree (*Arbutus unedo*) and giant butterwort (*Pinguicula grandiflora*) — are not found growing wild in mainland Britain. Others, such as the Cornish heath (*Erica vagans*) and the pale butterwort (*Pinguicula lusitanica*), are found in southwestern England as well as in Ireland. It therefore seems likely that these plants spread from Spain and Portugal up the Atlantic seaboard of Europe in postglacial times but were subsequently cut off by the rising sea levels. It is not impossible that some or all of the species may have survived the last glaciation in oceanic south-west Ireland, but direct proof of this is lacking.

Some mammals also failed to make the crossing into Ireland, and it is difficult to explain why they failed when closely related species succeeded. For example, the pygmy shrew reached Ireland, yet the common shrew did not (Fig. 11.11a and b). Perhaps this is related to their habitat preferences, for the pygmy shrew is found on moorlands and might have survived better than the common shrew if the conditions of the land bridge were of this type. The stoat also reached Ireland, but the weasel did not (Fig. 11.11c and d). Arrival in this case may have

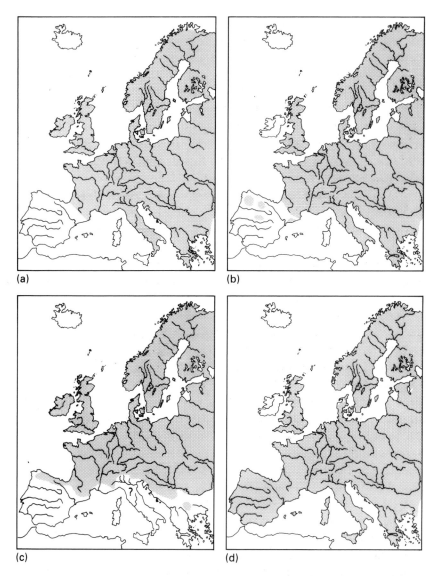

Fig. 11.11 Distribution maps of four mammal species in Europe. (a) Pygmy shrew (*Sorex minutus*), (b) common shrew (*Sorex araneus*), (c) stoat (*Mustella erminea*) and (d) weasel (*Mustela nivalis*). Of these, the pygmy shrew and the stoat reached Ireland, but the common shrew and the weasel failed.

been due to sheer chance. A single pregnant female arriving on a raft of floating vegetation could have been sufficient to populate the island. Discoveries in archaeological sites of small mammals like the wood mouse raise the possibility that some plants and animals could have been carried over the water by prehistoric humans. The arrival of some large mammals on isolated islands could also be an outcome of human transport, even in pre-agricultural times. Mesolithic people in the British

Isles, for example, may have been responsible for carrying the red deer (*Cervus elaphus*) to Ireland and to other offshore islands such as Shetland. The red deer was the major prey animal of these people, and the transport of young animals would have presented few difficulties even in small, primitive boats.

Rising sea levels during the present interglacial were responsible for the severing of land connections in many other parts of the world also. For example, Siberia and Alaska were connected across what are now the Bering Straits, which in places are only 50 miles (80 km) across and 50 m deep. This high-latitude land bridge would have been a suitable dispersal route only for arctic species, but it is believed to have been the route by which man entered the North American continent (see p. 288).

Time of warmth

The period of maximum warmth during the present interglacial lasted from about 7000 to 5000 years ago. At this time, warmth-demanding species extended farther north than they do at present. For example, the hazel (*Corylus avellana*) was found considerably further north in Sweden and Finland than it is today. This indicates that conditions have become cooler since that time. The remains of tree stumps, buried beneath peat deposits at high altitude on mountains and far north of the tree line in the Canadian Arctic, also bear witness to more favourable conditions in former times. Things are not always what they seem, however, and one has to remember the possible involvement of humans in the clearance of forests and the modification of habitats. Humans may have played an important part, for example, in the forest clearance that led to the formation of many of the so-called 'blanket mires' of western Europe [25].

In what is now the prairie region of central North America, coniferous woodland existed as recently as about 200 years ago. Such woodland may have been restricted to steep hillsides where the fires which swept across the rolling plains would not have burned with such severity.

In the Mediterranean region the extent of oak woodland was formerly greater than at the present day, but here the activities of humans have had a strong influence for many thousands of years, and climatic influences upon the vegetation are, therefore, difficult to discern.

The spread of forest, together with increasing warmth in the temperate regions, created unsuitable conditions for many of the plants which had previously been widespread at the close of the glacial stage. Some of these, the arctic−alpine species, are unsuited physiologically to high temperature. Many such plants, for example the mountain avens (*Dryas octopetala*), grow poorly when summer temperatures are high (above 23°C for Britain and 27°C for Scandinavia). The climatic changes which occurred during the post-glacial therefore proved harmful to such species, and many of them became restricted to higher altitudes, especially in lower latitudes. Other plant species are more tolerant of high temperatures but are incapable of survival under dense shade. Low-latitude, low-altitude habitats which became covered by forests were unsuitable for the continued growth of these species, and many of them also became restricted to mountains where competition from shade-casting trees and shrubs did not occur to the same extent. But for these species there were other opportunities. Lowland environments which for some reason bore no forest provided suitable places of refuge. Coastal dunes, river cliffs, habitats disturbed by periodic flooding, steep slopes, all provided sufficiently unstable conditions to hold back forest development and allow the survival of these plants.

The result of these processes was the production of relict distribution patterns. Sometimes the separation of species into scattered populations, even though it has lasted only about 10000 years, has permitted genetical divergence, as in the case of the desert pupfish (p. 269). For example, the sea plantain (*Plantago maritima*) has survived in both Alpine and coastal habitats, but the different selective pressures of the two environments have resulted in physiological divergence between the two races.

Some of those species, which were limited by competitive interactions rather than by climatic ones, have taken advantage of the disturbed conditions provided by human settlements and agriculture. These plants, which fared so poorly during forested times in the temperate latitudes, have become latter-day weeds and opportunists.

Climatic deterioration

Although there has been an overall cooling of the climate during the past 5000 years, this has not taken place in a gradual way, but in a series of steps. Beginning about 3000 BC, a number of quite sudden temperature falls in the Northern Hemisphere had the effect of halting the retreat of glaciers and of increasing the rates of bog growth. One of the most pronounced of these steps, as far as north-west Europe is concerned, occurred about 500 BC and caused a sudden increase in the rate of growth of bogs over the entire area. In many bogs this has left a permanent mark upon the stratigraphic profile of the bog; a dark, oxidized peat typical of slow-growing bog surfaces is suddenly replaced by the almost undecomposed vegetable matter that typifies a fast-growing bog. The German botanists who first described this phenomenon called it the *Grenzhorizont* (boundary horizon) and the name is still frequently used by palaeobotanists. Using such evidence of increased bog growth as an indicator of wetter or cooler climate is, however, fraught with problems. Often such changes are of only local significance and are probably associated with local drainage patterns, human land use, or peculiarities of microclimate or bog hydrology. Only those changes which are synchronous over large areas can be considered as truly climatically induced.

Data relating to changes in bog growth rate over northern Europe are summarized in Fig. 11.12. It is difficult to reach firm conclusions from these data, but there is some consistency in the increased wetness (or

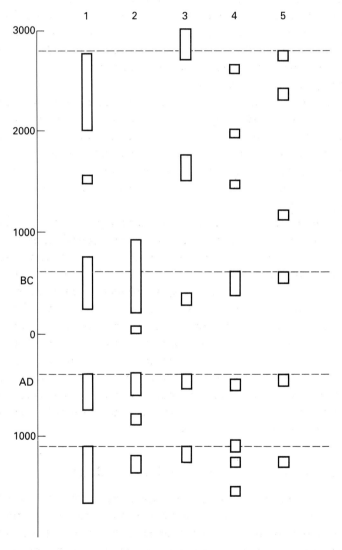

Fig. 11.12 Periods of active raised bog growth in northern Europe, which may correlate with episodes of wetter and/or cooler climate. Key to columns: (1) Ireland, (2) Britain, (3) Netherlands and Germany, (4) Denmark, (5) Scandinavia. Dates are in uncorrected radiocarbon years. Dotted lines represent times of fairly general rejuvenation of bogs.

lower temperature, or both) found in the early centuries BC and after about AD 400.

More reliable information has been obtained from analyses of oxygen isotopes in cores collected from the Greenland ice-cap (Fig. 11.13) [26]. Snow and ice deposited under warm conditions have higher proportions of the isotope ^{18}O than those accumulating in the cold (see p. 255). The changing proportions of isotopes in the ice cap therefore provide a detailed record of fluctuating temperature. In Fig. 11.13 the oxygen isotope record is compared with curves from Iceland and England which have been compiled from various types of indirect evidence, such as early literary sources. The combination of these pieces of information indicates that there have been considerable variations in climatic patterns even during recent, historical times.

Evidence of climatic change derived from the growth rings of trees also serves to emphasize regional variations in climate. For example, work on tree rings in pine from Fennoscandia, forming a continuous record back to AD 500, shows a great deal of variation and suggests that the so-called Little Ice Age in that region was a relatively brief event, lasting only from 1570 until 1650. In England it lasted from 1300 to 1700 (Fig. 11.13).

Some biological consequences of the generally deteriorating climate over the past 5000 years are apparent from changing distribution patterns. In Europe, such species as the hazel (*Corylus avellana*) extended further north and up to higher altitudes in Scandinavia than they do today, and the pond terrapin (*Emys orbicularis*) had a wider occurrence in north-western Europe. It is very difficult, however, to be sure that such contractions in range have not resulted from human destruction of either the species or their habitats during this time period, because the impact of human cultures on the natural landscape through these millennia has become increasingly severe. In North America one might regard the resurgence of spruce in the north-west (Fig. 11.4a) as a product of climatic deterioration and an indication that the current interglacial is entering its latter days.

The environmental impact of early human cultures

As the postglacial climatic optimum passed and conditions in the temperate regions became generally cooler and wetter, the agricultural concept continued to spread into higher latitudes, and with agriculture came the incentive to modify the environment to make it more suitable for enhanced productivity of domestic animals and plants. Temperate forests are unsuitable for the growth of domesticated plants, most of which have a southern origin and a high demand for light. Similarly, domestic animals such as sheep and goat are not at their most efficient

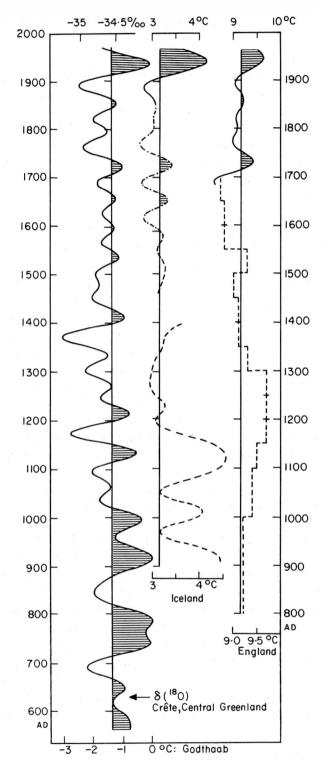

Fig. 11.13 Oxygen isotope curve from an ice core taken in central Greenland (left), and the projected temperature curve for Iceland (left) and for England (right). From Dansgaard *et al.* [26].

in a woodland habitat, preferring open grassland conditions. Cattle and pigs, on the other hand, can be herded within forest, but even they can be managed more efficiently in a habitat that is more open. Even in pre-agricultural times the Mesolithic people of northern Europe had discovered that the opening of forest and burning to retain open glades provided a higher productivity of red deer. Like some of the palaeoindian tribes of North America, who became closely dependent on the bison, the European people of the Middle Stone Age were often reliant on the red deer.

The intensification of forest clearance with the coming of agriculture in northern Europe is very apparent from the pollen diagrams (see Fig. 11.3), where the pollen of open habitat species (such as grasses, plantains and heathers) rises and the proportion of tree pollen falls. The precise pattern of forest clearance and the development of heathland, grassland, moorland and blanket bog as consequences of this activity varies from one area to another, depending on the local conditions and the pattern of human settlement. By 2000 years ago the impact was severe through much of central and western Europe, although the forests of the far north had been little influenced by mankind at that time. Some of the most severe deforestation, judging from the pollen record, had taken place in the north-west of Europe, including the British Isles. Perhaps it was in this region that the forest was least able to recover from human impact, and the maintenance of heavy grazing kept the area relatively open.

In North America, agriculture in the temperate zone in pre-European times was confined largely to the growing of maize and other weedy species, including purslane (*Portulaca oleracea*). This involved the clearance of small areas of forest, and the effect of these clearings can be detected in pollen diagrams [27]. Although these clearings seemed to recover, and there is little evidence for large-scale forest destruction of the European type, there are changes in the composition of the North American forests that may well have resulted from the activities of agricultural peoples. The burning and cutting of forests is often associated with a loss in certain species, such as sugar maple and beech, and an increase in the abundance of fire-resistant pines and oaks, together with a general increase in the frequency of birch. Intensive clearance in the eastern United States and Canada was usually delayed until the arrival of European settlers in the eighteenth and nineteenth centuries.

Recorded history

As soon as human beings appeared on the scene, they often inadvertently began to leave information about climate and its changes. Early records are clues rather than precise information, such as the ancient rock drawings of hunting scenes discovered in the Sahara, which indicate that its climate was much less arid at the time they were made than it is today. With the development of writing, accurate records of climatic changes began to be made. For example there are records of pack ice in the Arctic seas near Iceland in 325 BC, indicating the very low winter temperatures at the time. During the heyday of the Roman Empire, however, there was steady improvement in climate, allowing the growth of such crops as grapes (*Vitis vinifera*) and hemp (*Cannabis sativa*) even in such relatively bleak outposts as the British Isles. This warmer, more stable period reached its optimum between AD 1000 and 1250, after which it again grew colder.

In 1250 alpine glaciers grew and pack ice advanced in the Arctic seas to its most southerly position for 10000 years. In 1315 a series of poor summers began in northern Europe, and these led to crop failures and famine. Climatic deterioration continued and culminated in the 'Little Ice Age' of AD 1300–1700, during which the glaciers reached their most advanced positions since the end of the Pleistocene glacial epoch [28]. During this time, trees on the central European mountains were unable to grow at their former altitudes, due to the increasingly cold conditions. The climate has become warmer between 1700 and the middle of the present century, since when it has again grown cooler. After 1940 winters became colder, and since 1970 average temperatures have been rising again.

During historic times all changes in plant and animal distributions have probably been affected to some extent by humankind and, as far as climatic change is concerned, we must be cautious in our interpretation of them. Because of the complexity of human influence on the environment, it is difficult to tell whether some recent changes in the distribution of certain organisms have been influenced by climate. It is possible that some have, such as that of the lake-dwelling holly-leaved naiad (*Naias marina*), which was fairly widespread in the British Isles during the early postglacial warm period, but is now virtually extinct, as a direct result of the overall fall in temperature, surviving only in a few localities in Norfolk. A further example is the lizard orchid (*Himantoglossum hircinum*), a scarce and beautiful plant of south-east England. During the first half of this century it extended its range considerably, but this has contracted again since 1940, possibly as a response to the colder climate since that time. All orchids are extremely sensitive to changes in their environment, and many species may have their distribution limited by minor climatic fluctuations.

In the last decade average world temperatures have been rising again, so it remains to be seen whether a response will be observed in the distribution patterns of certain plants and animals.

Current and future changes

The current trend in global temperature is distinctly upwards (Fig. 11.14). The trend began in the mid-nineteenth century, but was interrupted between 1940 and 1970 when a plateau, or even a slight reversal, was experienced. But now the curve is rising steeply and is consequently a cause of some concern. In a world where climatic fluctuation is commonplace, it is difficult to ascribe any given wiggle in the curve to a specific cause. The current rise, however, is clearly a strong and rapid one, so the possibility that it has been generated by human activity over the past century or so must be considered.

The possible role of the atmosphere in this process has received particular attention, especially the accumulation of certain gases that enhance the heat retention properties of the atmosphere and create a 'greenhouse effect'. Gases such as carbon dioxide, methane, nitrogen oxides, water vapour and the chlorofluorocarbons (CFCs) all absorb energy strongly in the infra-red (long wavelength) part of the spectrum. Energy from the sun penetrates the atmosphere as short wavelength light, but is converted in part to long wavelength energy on striking the surface of the planet. If the escape of this energy by radiation is prevented as a result of absorption by the thermal blanket of the atmosphere, then the temperature of the earth may be expected to rise.

The carbon dioxide content of the atmosphere has been accurately monitored in the Hawaiian station of Mauna Loa, far from the influence of any local industrial activity, since 1958. The average CO_2 content of the atmosphere was then 315 parts per million (0·0315 per cent). In 1983 the level had risen to 341 p.p.m. (see Fig. 11.15). It is believed that the level of carbon dioxide has been rising since the beginnings of intensive

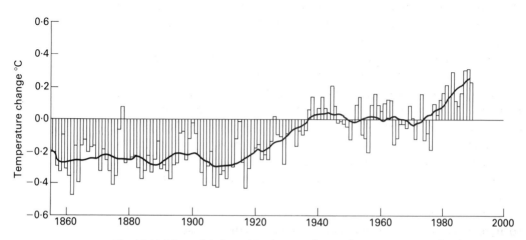

Fig. 11.14 Mean global combined sea surface and air temperatures for the past 130 years relative to the average for the period 1951–1980.

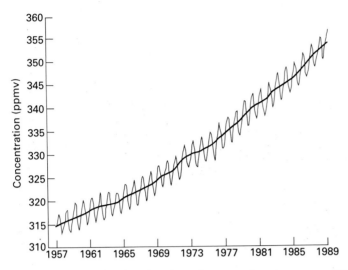

Fig. 11.15 Concentration of carbon dioxide in the atmosphere (parts per million by volume) over the past 34 years as measured in Hawaii.

global forest clearance for agriculture [29], certainly over the last 200 years, from a base level of about 280 p.p.m., and the industrial output has been accelerating since 1860. It is probable that the carbon entering the atmosphere from forest clearance exceeded that from the burning of fossil fuels until 1960, since which there has been approximately a 250 per cent increase in fossil fuel burning.

Currently, the annual emission of carbon dioxide from industrial activities (largely fossil fuel burning) is estimated at 6 Gt carbon. The geographical origins of these emissions are shown in Fig. 11.16 and it can be clearly seen that the industrialized regions of the Northern Hemisphere are the main sources. Land-use changes, including tropical forest clearance, are estimated to produce a further 1−2 Gt carbon per annum. Some of this extra carbon entering the atmosphere is removed by solution in the oceans, or by the photosynthesis of intact vegetation (but this only represents a net uptake if the vegetation is growing in biomass). Despite these sinks for carbon, the atmospheric load of the element is currently increasing by 3·4 Gt per annum, and thus CO_2 constitutes the most important contributor to the greenhouse effect.

Methane is present in much smaller quantities in the atmosphere (less than 2 p.p.m. compared with 355 p.p.m. for carbon dioxide) but it is a more efficient agent for the absorption of infra-red radiation (about 3·7 times that of carbon dioxide), so its presence is significant. Its atmospheric concentration has only been monitored globally since 1984, but during the first 5 years of recording it increased by about 4 per cent. This represents an accumulation rate of about 4·4 million tons per annum. Most of this comes from the world's wetlands, both natural and artificial

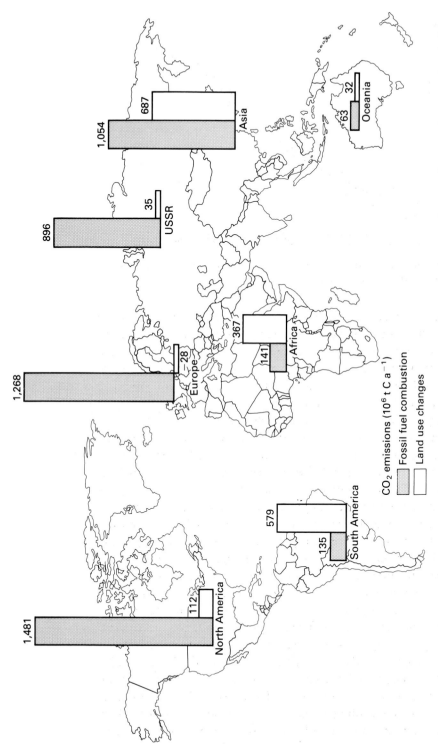

Fig. 11.16 Emissions of carbon dioxide to the atmosphere as a result of human activities during 1980. Two types of emission are recognized, those resulting from the burning of fossil fuels and those resulting from land-use change (e.g. forest clearance). The negative value of the latter for Europe implies that this continent acted as a carbon sink during the year. From UNEP [29].

(such as rice paddies), and also from the exhalations of ruminant animals and the activity of termites, together with such human activities as gas drilling, biomass burning and landfill operations. The low latitudes are the main source areas for methane, both because of tropical wetlands and the abundance of rice paddies in the tropics (see Fig. 11.17). But the high-latitude peatlands are also an important natural source of this greenhouse gas.

The chlorofluorocarbons (or halocarbons) are synthetic chemicals used in refrigerators, polystyrene manufacture and in aerosol propellants, and they are important both as greenhouse gases and as potential destroyers of stratospheric ozone (a gas that shields the earth from much of the harmful ultraviolet rays from the sun). Their concentration in the atmosphere has increased by about 100 per cent over the past 20 years.

The sum of these various atmospheric pollutants may well account for the change in global temperature over the past century but the case, although likely, is not fully proven. It is even more difficult to predict the possible outcome of climatic change in terms of global vegetation, species distributions and human agricultural constraints should it continue at its current rate. If the carbon dioxide level were to increase to, say, 660 p.p.m., the global rise in temperature would be about $3 \pm 1.5°C$. But this does not take into account possible feedback effects, such as

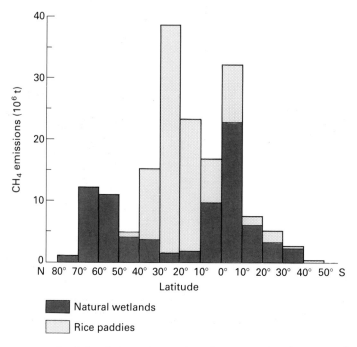

Fig. 11.17 Latitudinal distribution of annual methane emissions from natural wetlands and rice paddies. From UNEP [29].

increased evaporation from the oceans causing increased cloud cover and hence reduced solar input to the earth's surface. Rising temperature would also cause melting of the glaciers and ice caps, resulting in raised sea levels and also changes in the earth's reflective properties (albedo). There are thus many imponderables and no simple answers. Several global climatic models have been produced and many possible scenarios tested to predict climatic changes, and these have been applied to questions of species ranges and changes in the composition and geographical limits of the major biomes.

A warmer world should see a greater potential for plant productivity (since carbon dioxide is often the limiting factor for current photosynthesis), together with a movement of many species towards higher latitudes and a withdrawal from lower latitudes. Such is the prediction, for example for the beech tree (*Fagus grandifolia*) in North America [30] (Fig. 11.18). Two possible scenarios are considered, but both show a northward shift in distribution pattern in response to higher temperature. At the other end of the process, the retreat of species from areas where they have formerly succeeded is already apparent in some areas. The red spruce (*Picea rubens*), for example, is declining over much of its range in the eastern United States, and is believed to have been in decline

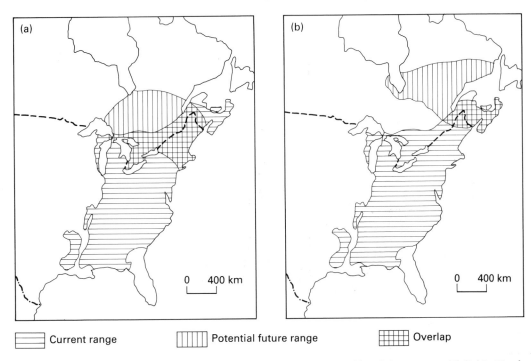

Fig. 11.18 Projected change in the distribution of beech (*Fagus grandifolia*) in North America under two greenhouse scenarios; (a) less severe and (b) more severe. From Roberts [30].

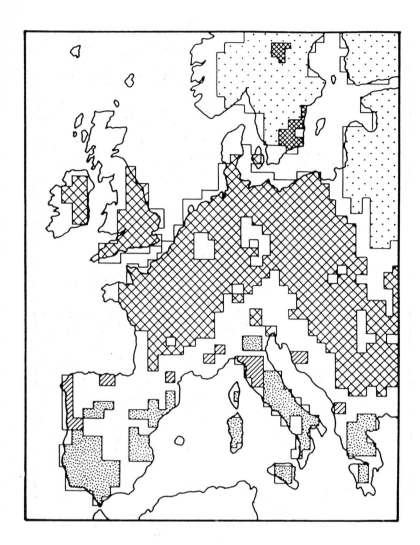

Ecosystem types

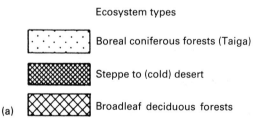

Boreal coniferous forests (Taiga)

Steppe to (cold) desert

(a)   Broadleaf deciduous forests

Fig. 11.19 Model of the potential natural vegetation of Europe based on (a) the current average temperature and precipitation and (b) an increase of + 5°C in average temperature together with + 10 per cent average precipitation. Note the northerly shift in vegetation zones and the increase in the area of temperate evergreen forest. From de Groot [32].

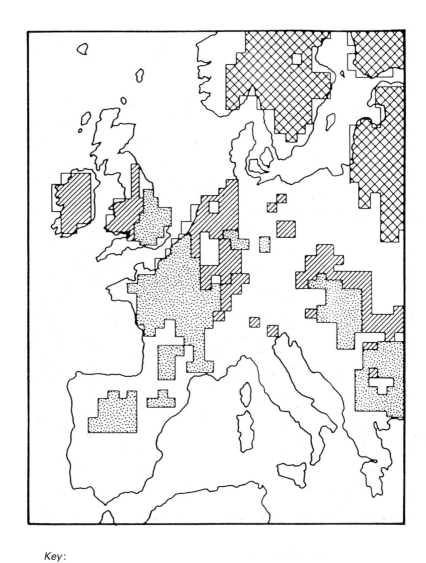

Key:

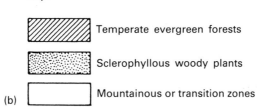

Temperate evergreen forests

Sclerophyllous woody plants

Mountainous or transition zones

(b)

since the early part of last century. It has been proposed [31] that this change is a response to a general increase in both mean annual and summer temperatures.

The shift of many such species will lead to a change in the distributions of the earth's biomes (see Chapter 4). There have been several attempts to model the changes that would be brought about by a range of possible climatic changes, and one of them is shown in Fig. 11.19 [32]. Here the current state of affairs in Europe is compared with predicted vegetation characteristics given a rise of 5°C together with an increase in average precipitation of 10 per cent. Effectively, the boreal forests of Scandinavia would be replaced by deciduous forest, and the zone currently bearing deciduous forest would develop a Mediterranean type of vegetation. In practice it is questionable whether these changes could take place because of the rapid rate of climatic change, the relatively slow response rate of vegetation (due to slow breeding and spread, especially among trees) and the fact that so much of the available land area is already greatly modified by human activity. Perhaps the most immediate response would be seen among invertebrate animals, especially those with the capacity for flight (compare the beetle response at the end of the last glaciation – p. 280), and among those plants that have a short generation time and an efficient system of seed dispersal (i.e. weed species).

Not only would wild plant and animal species be forced into higher latitudes with an increase in global temperature, but the influence would also be felt by domesticated species. At present maize can be grown in Britain only as a silage crop for the production of cattle food. Grain maturation is poor in the relatively cool moist summers of the area, because ripening requires 850 degree-days above a base temperature of 10°C (number of degrees above 10°C multiplied by the number of days experiencing these temperatures). If the average temperature were to rise by 3°C, only the extreme north of Scotland would be unsuitable for growing grain maize. Similar changes in the United States (an increase in temperature of 3°C and precipitation of 8 cm) would cause a shift in the grain belt to the north-east, around the Great Lakes [33].

Not only grain production and maturation, but also grain quality could be affected by rising temperatures. Some research from Australia shows that, when wheat grain is produced at high temperature (above 30°C), it is less suitable for the production of bread because the strength of the dough is reduced. This could have serious implications for wheat-growing in countries such as Australia where high summer temperatures are to be expected in the greenhouse world.

Alterations in the pattern of precipitation over the surface of the earth are particularly difficult to predict, but some parts of the subtropics and beyond are likely to become yet drier in a greenhouse world. Many of the areas currently threatened by desertification will fare even worse in such circumstances, with severe social consequences for human populations.

The future undoubtedly holds changes for the earth's biosphere, only some of which are predictable. Perhaps it is the unpredictable ones which are most to be feared. The changes in climate and in the physical and chemical conditions of our planet will result in a modification of its biogeography, inevitably involving the extinction of some species. The study of the requirements and the interactions of plants and animals, including their past distribution patterns and evolution, provides an important means of understanding the complexity of nature and also of our own role within it. The story of our planet has been one of constant change and adjustment, and the biosphere has proved remarkably resilient to all the changes so far experienced. It may seem complacent to have faith in the ability of the natural world to cope with any predictable stresses, including those directed at it by our own species, but the probability is that this is the case. The most important question facing our own species is whether we are equally resilient, and whether we are equally able to adapt to changing conditions fast enough to permit the survival of *Homo sapiens*. Nature, in some form, will undoubtedly survive the next few centuries, but will we?

References

1 Broecker, W.S., Kennett, J.P., Flower, B.P., Teller, J.T., Trumbo, S., Bonani, G. & Wolfli, W. (1989) Routing of meltwater from the Laurentide Ice Sheet during the Younger Dryas cold episode. *Nature* **341**,318–321.

2 Jones, G.A. (1991) A stop–start ocean conveyer. *Nature* **349**,364–365.

3 Engstrom, D.R., Hansen, B.C.S. & Wright, H.E. Jr (1990) A possible Younger Dryas record in southeastern Alaska. *Science* **250**,1383–1385.

4 Kudrass, H.R., Erlenkeuser, H., Vollbrecht, R. & Weiss, W. (1991) Global nature of the Younger Dryas cooling event inferred from oxygen isotope data from Sulu Sea cores. *Nature* **349**,406–409.

5 Fairbanks, R.G. (1989) A 17 000-year glacio-eustatic sea level record: influence of glacial melting rates on the Younger Dryas event and deep-ocean circulation. *Nature* **342**,637–642.

6 Dansgaard, W., White, J.W.C. & Johnsen, S.J. (1989) The abrupt termination of the Younger Dryas climate event. *Nature* **339**,532–534.

7 Bennett, K.D. (1983) Postglacial population expansion of forest trees in Norfolk, U.K. *Nature* **303**,164–167.

8 Jacobson, G.L., Webb, T. & Grimm, E.C. (1987) Patterns and rates of change during the deglaciation of eastern North America. In *The Geology of North America*, Volume K-3 (eds W.F. Ruddiman & H.E. Wright). Geological Society of America, New York, pp. 277–288.

9 Pilbeam, D. (1984) The descent of hominoids and hominids. *Scientific American* **250**(3),60–69.

10 Hay, R.L. & Leakey, M.D. (1982) The fossil footprints of Laetoli. *Scientific American* **246**(2),38–45.

11 Andrews, P. (1984) The descent of man. *New Scientist* **102**(1408),24–25.

12 Jelinek, A.J. (1982) The Tabun Cave and Paleolithic man in the Levant. *Science* **216,**1369–1375.

13 Rogers, R.A., Rogers, L.A., Hoffmann, R.S. & Martin, L.D. (1991) Native American biological diversity and the biogeographic influence of Ice Age refugia. *Journal of Biogeography* **18,**623–630.

14 Martin, P.S. & Wright, H.E. Jr (1967) *Pleistocene Extinctions: the Search for a Cause.* Yale University Press, New Haven.

15 Grayson, J.E. (1977) Pleistocene avifaunas and the overkill hypothesis. *Science* **195,**691–693.

16 Stuart, A.J. (1991) Mammalian extinctions in the Late Pleistocene of Northern Eurasia and North America. *Biological Reviews* **66,**453–562.

17 Lewin, R. (1983) What killed the giant mammals? *Science* **221,**1036–1037.

18 Loy, T.H. (1983) Prehistoric blood residues: detection on tool surfaces and identification of species of origin. *Science* **220,**1269–1271.

19 Hubbard, R.N.L.B. (1980) Development of agriculture in Europe and the Near East: evidence from quantitative studies. *Economic Botany* **34,**51–67.

20 Van Der Merwe, N.J. (1982) Carbon isotopes, photosynthesis and archaeology. *American Scientist* **70,**596–606.

21 Iltis, H.H. (1983) From teosinte to maize: the catastrophic sexual transmutation. *Science* **220,**886–894.

22 Davis, S. (1982) The taming of the few. *New Scientist* **95,**697–700.

23 Rosignol-Strick, M., Nesteroff, W., Olive, P. & Vergnaud-Grazzini, C. (1982) After the deluge: Mediterranean stagnation and sapropel formation. *Nature* **295,**105–110.

24 Lezine, A.M. (1989) Late Quaternary vegetation and climate in the Sahel. *Quaternary Research* **32,**317–334.

25 Moore, P.D., Merryfield, D.L. & Price, M.D.R. (1984) The vegetation and development of blanket mires. In *European Mires* (ed. P.D. Moore). Academic Press, London, pp. 203–235.

26 Dansgaard, W., Johnsen, S.J., Reeh, N., Gundestrup, N., Clausen, H.B. & Hammer, C.U. (1975) Climate changes. Norsemen and modern man. *Nature* **225,**24–28.

27 McAndrews, J.H. (1988) Human disturbance of North American forests and grasslands: the fossil record. In *Vegetation History* (eds B. Huntley & T. Webb III). Kluwer, Dordrecht, pp. 673–697.

28 Crowley, T.J. & North, G.R. (1991) *Paleoclimatology.* Oxford University Press, New York.

29 UNEP (1991) *Environmental Data Report,* 3rd edn. Basil Blackwell, Oxford.

30 Roberts, L. (1989) How fast can trees migrate? *Science* **243,**735–737.

31 Hamburg, S.P. & Cogbill, C.V. (1988) Historical decline of red spruce populations and climatic warming. *Nature* **331,**428–431.

32 De Groot, R.S. (1987) *Assessments of potential shifts in Europe's natural vegetation due to climatic change and implications for conservation.* Young Scientist's Summer Program 1987: Final Report. International Institute for Applied Systems Analysis. Luxemburg, Austria.

33 Parry, M. (1990) *Climate Change and World Agriculture.* Earthscan, London.

Index

Page references in *italic* type indicate figures or tables.